KU-548-473

This Book
belongs to the Library of
King Edward VI's
Grammar School,
Guildford, Surrey.

Remarkable Discoveries!.

The personalities, the triumphs, the trials and tribulations are all here in this fascinating book. *Remarkable Discoveries!* takes the reader on an exhilarating tour through some of the major scientific discoveries that have benefited humanity. Exciting discoveries are found in all areas of science. Physics, cosmology, biology, medicine, chemistry, geophysics and mathematics are all represented.

Written for the reader with little knowledge of science, the book entertains and enthuses. There is a common theme of the element of surprise and wonder that accompanies many scientific discoveries.

The book is engagingly written and leaves the reader eager to know more about the world of science and scientists.

Frank Ashall was born in Bradford, Yorkshire, in 1957 and obtained his bachelor's degree in biochemistry from Oxford University in 1980. He did his doctoral research in the Sir William Dunn School of Pathology, Oxford, under the supervision of Professor Sir Henry Harris. After spending three years studying the molecular basis of cancer in Dr Theodore T. Puck's laboratory at the Eleanor Roosevelt Institute for Cancer Research, Denver, Colorado, he returned to Britain, where he worked on tropical diseases, first at the London School of Hygiene and Tropical Medicine, then at Imperial College, London. Currently, he is an assistant professor in the Psychiatry Department at Washington University School of Medicine, St Louis, where he is investigating the molecular basis of Alzheimer's disease. Dr Ashall worked with the British newspaper, *The Independent*, in 1991 as a Media Fellow of the British Association for the Advancement of Science and has since written numerous articles on popular science. He has also written several poems on science and medicine and is a member of the International Society of Poets. He believes fervently that scientists have a duty to inform the public about the wonders of Nature and its beneficial applications to society.

Also of interest in popular science

Hard to Swallow: a brief history of food
RICHARD LACEY
An Inventor in the Garden of Eden
ERIC LAITHWAITE
The Outer Reaches of Life
JOHN POSTGATE
Prometheus Bound
JOHN ZIMAN

FRANK ASHALL

R.G.S. GUILDFORD,
J. C. MALLISON
LIBRARY.

Remarkable Discoveries!

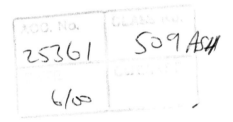

ACC. No.
25361

CLASS No.
509 ASH

6/00

CAMBRIDGE
UNIVERSITY PRESS

Published by the Press Syndicate of the University of Cambridge
The Pitt Building, Trumpington Street, Cambridge CB2 1RP
40 West 20th Street, New York, NY 10011-4211, USA
10 Stamford Road, Oakleigh, Melbourne 3166, Australia

© Cambridge University Press 1994

First published 1994

Printed in Great Britain by
Biddles Ltd, Guildford, Surrey

A catalogue record for this book is available from the British Library

Library of Congress cataloguing in publication data
Ashall, Frank, 1957–
Remarkable Discoveries! / Frank Ashall.
p.   cm.
ISBN 0 521 43317 7 (hc)
1. Discoveries in science.   I. Title.
Q180.55.D57A78   1994
509–dc20   93-46796 CIP

ISBN 0 521 43317 7 hardback

To my daughter, Etty,
and all children of her generation

Contents

Preface

Scientists have a responsibility to inform the public about the good that basic scientific research has done and can do for humanity, and to educate everyone about the wonderful workings of Nature. An extremely important point to get across, in my view, is that nobody can predict what benefits will come from pure research: basic studies aimed at understanding Nature have, time and time again, led to unanticipated applications of enormous significance. I hope this book goes some way towards addressing this issue.

In recent years scientific research has become highly commercialised, even in the academic laboratories of our great universities, many of whose ideals regarding freedom of pursuit of knowledge for its own sake have been tarnished. Government funding of pure scientific research is pitifully poor world-wide and it is disgracefully dwarfed by the tremendous financial support that is available for military research. Even when funding for basic research is forthcoming, some areas of gigantic human significance still fail to receive the funding that they deserve, not to mention those research programmes that have to be aborted because they are grossly underfunded. Who can put a price on the curiosity of Michael Faraday, which gave rise to almost the whole of today's electrical industry? What financial worth were the thoughts of Louis Pasteur, whose insatiable desire to interpret Nature contributed much of today's understanding, prevention and treatment of infectious diseases?

The organisations that fund scientific research should continually consider discoveries of the type described in this book and bear in mind the immense wisdom of funding pure research, whatever the subject. The key lesson to be learned from scientific discoveries of the past is that basic investigation of Nature will not only add to knowledge about the marvellous and beautiful Universe in which we live, but also, inevitably, it will give us new and unexpected benefits that will improve every aspect of our daily lives. The beneficial applications of science are for everyone, everywhere, to enjoy, and the beauty of the laws and intricacies of Nature are for all to ponder with awe. Sadly, members of the governments of our world generally lack adequate scientific training, and consequently they fail to realise how important scientific research is, and, tragically, hundreds of millions of innocent people of poor countries still do not enjoy the benefits of most of the discoveries described in this book.

In 1991 I spent two months as a Media Fellow (a research scientist working with the media) with the British newspaper, *The Independent*, where my interest in areas of science outside the scope of my own specific research topic was stimulated. It was there that I realised how little many research scientists know about areas of science beyond their own specialty, and it was there that the tremendous importance of communicating science to the public was brought home to me. I am truly indebted to Dr Tom Wilkie, Science Editor of *The Independent*, for providing the stimulation that spurred me to write this book and for giving me the opportunity to work with the newspaper. Others at *The Independent*, particularly Susan Watts and Steven Connor, were also helpful during my time as a Media Fellow. The Committee on the Public Understanding of Science (COPUS), in conjunction with the British Association for the Advancement of Science, sponsored the Media Fellowship and I am grateful to them also.

I thank Dr Robert Harington and Dr Jo Clegg at Cambridge University Press, for their highly valuable and constructive comments. My wife, Dr Alison Goate, also provided helpful comments on the manuscript. The following people contributed very useful information and helpfully critical comments, and I am grateful to them: Professor Alec Jeffreys, University of Leicester; Professor Harold Kroto, University of Sussex; Professor Cesar Milstein, University of Cambridge; Professor Sir Edward Abraham, Sir William Dunn School of Pathology, University of

Oxford; Professor Sir Henry Harris, Sir William Dunn School of Pathology, University of Oxford; Professor Clifford Will, Washington University in St Louis, Missouri; and Professor Wai-Mo Suen, Washington University in St Louis, Missouri.

Frank Ashall, Hampton Wick, England

Useful information

1 kilometre (km) = 1000 metres (m) = 0.621 miles (mi)
1 mi = 1.61 km
1 centimetre (cm) = 0.394 (in)
1 in = 2.54 cm
1 foot = 12 in = 0.305 m

1 (UK) pint = 0.568 litres
1 (US) pint = 0.454 litres

1 kilogram (kg) = 1000 grams (g) = 2.20 pounds (lb)
1 lb = 0.454 kg
1 ton = 2240 lb = 1016 kg

1 million = 1 000 000 = 10^6
1 billion = 1 000 000 000 = 10^9 (US terminology)
1 trillion = 1 000 000 000 000 = 10^{12} (US terminology)

Speed of light in a vacuum (c) = 300 000 kilometres per second = 186 000 miles per second.

To convert from degrees Centigrade (Celsius) to degrees Fahrenheit, multiply by 9, then divide by 5, then add 32 degrees.

0 °C = 32 °F
100 °C = 212 °F

1
The Father of Electricity

In 1991 the Science Museum in London held a special exhibition to celebrate the bicentenary of the birth of Michael Faraday. At the entrance to the exhibition stood a statue of Faraday surrounded by more than a dozen household items, including a vacuum cleaner, an electric sewing machine, a hairdryer and a food mixer. Beneath the statue a plaque read, 'All the electrical equipment we use today depends upon the fundamental discoveries made by Michael Faraday, 1791–1867.' This was no exaggeration: Faraday's experiments into electricity and magnetism form the basis of the whole electrical industry today and modern society can be grateful to Faraday for many of its luxuries and life-saving facilities. This British scientist, who had no formal university education, truly deserves to be called the Father of Electricity.

Yet the wonderful and diverse applications that eventually developed from Faraday's work were not foreseen by Faraday himself when he began his researches, and in no way did he directly seek applications of his work. His aims were to investigate Nature for its own sake – to reveal the beauty of the physical world through experimentation. If the work of any scientist quintessentially exemplifies Louis Pasteur's dictum that 'there is no such thing as applied science, only the application of pure science', Faraday is that scientist. Through his deep passion and insatiable desire for pure knowledge, he revealed facets of physics that have revolutionised our planet.

Historical context

To understand the significance of Michael Faraday's discoveries and how they came about, we need to assess the state of knowledge about electricity and magnetism in the early nineteenth century, when he began his experiments. The time was ripe for great discoveries in electricity and magnetism. Excellent progress had already been made in the fields of optics and mechanics during the seventeenth and eighteenth centuries, and electricity and magnetism were fast becoming fashionable areas to study when Faraday came onto the scientific scene.

The ancient Greeks undoubtedly knew about magnetism, particularly through the studies of the philosopher, Thales of Miletus, who lived during the sixth century BC. Thales demonstrated that a lump of iron ore (also called lodestone or magnetite) could attract iron towards it. Because the lodestone he used in his experiments came from the Aegean town of Magnesia, Thales called it 'Magnesian rock'. Hence, the word 'magnet'.

The Greeks were also familiar with electrostatic attraction. When a piece of amber (fossilised tree sap) was rubbed, it was able to attract light objects such as feathers. The term, 'electricity', comes from the Greek word *elektron* meaning amber. The Greeks were also aware of the similarities between magnetic and electrostatic attraction, although magnetic attraction was thought to be the more powerful while electrostatic attraction was considered to be the more versatile. This was because amber, when rubbed, could pick up many light objects regardless of the material from which they were made, whereas lodestone was able to attract only pieces of iron or other lumps of lodestone.

Progress in understanding magnetism was somewhat slow, but by the twelfth century **magnetic induction**, the process by which a lump of magnetic iron is able to cause another piece of non-magnetic iron to become a magnet itself, was known to exist. Most people today have demonstrated magnetic induction at some time in their life by stroking a piece of iron or steel (for example, a paper clip or a needle) with a magnet, with the result that the iron or steel becomes magnetised.

A magnet, when floating on water, points in a north–south direction. When it is made to point in any other direction, it returns to the north–south orientation. Magnets were shown to have a north and south pole, and two magnets attracted each other if their poles were oriented in an unlike fashion (south-to-north), whereas repulsion occurred if like poles (north-to-north or south-to-south) were placed together. This led to the rule: 'Like poles repel, unlike poles attract', and formed the basis of the magnetic compass, in which a magnetic needle aligns itself with the Earth's magnetic poles. Navigators were now able to orient themselves without using the Sun or Pole Star to guide them. The compass may have first been used by the Chinese. By the fifteenth century, great strides had been made in human exploration of the planet, thanks particularly to the magnetic compass.

William Gilbert (1544–1603), who was physician to Queen Elizabeth I of England, was one of the first scientists to investigate thoroughly and systematically the scientific basis of magnetism. He demonstrated elegantly that a compass needle not only points in a north–south direction but also dips downwards. He shaped a piece of magnetic lodestone into the form of a sphere to mimic the Earth and discovered that a compass needle pointed in one direction only and also dipped when it was placed near to the sphere. Gilbert interpreted this result as meaning that the Earth was itself an enormous magnetic sphere, and that it possessed a north and a south pole. In other words, a spherical magnet has polarity. This dispelled older ideas that somewhere in the Earth's distant north there was a gigantic mountain range made of iron or lodestone.

Gilbert also carried out experiments into the fundamentals of electrostatics. He showed that many precious gems, such as sapphires and diamonds, were, like amber, capable of electrostatic attraction when they were rubbed. He coined the word, 'electrics' to describe these substances.

Static electricity produced by rubbing electric substances was not a particularly abundant source of large amounts of electricity and electricity that was stored up in electrostatically charged materials was not easy to harness in a controlled manner: electrostatic discharge was always too rapid. What was needed in order to study electricity properly was for a system to be developed in which large amounts of electricity could be obtained and for this electricity to be harnessed over controllable time scales that would allow scientists to study its properties. The first major step towards solving this problem came from the German physicist, Otto von Guericke (1602–1686), who made a melon-sized ball of sulphur, which was a particularly good material for producing static electricity. When this ball of sulphur was rotated on a crank whilst touching another material, a large store of static electricity accumulated in the sphere. Von Guericke also discovered that electrostatically charged spheres could attract or repel each other, similarly to magnetic poles. He demonstrated that one sulphur sphere could induce another sphere to be electrostatically charged. This was called **electrostatic induction**. By the late seventeenth century, a growing number of scientists were beginning to think that electricity and magnetism were somehow closely related to each other.

During the eighteenth century and early nineteenth century, scientists were able to demonstrate that electricity could be made to flow through various materials, for example metal rods. The US scientist, Benjamin Franklin (1706–1790), proposed that electricity was a fluid that flowed from regions of positive charge to regions of a more negative charge. Nowadays we know that electricity involves a flow of negatively charged electrons from negative to positive poles of an electric circuit, that is, in the opposite direction to that proposed by Franklin.

Three major developments took place during the seventeenth and early eighteenth centuries that were to be particularly important for Faraday's subsequent discoveries: the first electric battery was invented; an instrument for measuring electric currents (a galvanometer) was developed; and a direct relationship between electricity and magnetism (electromagnetism) was discovered.

The electric battery was invented in 1800 by the Italian scientist, Alessandro Volta (1745–1827), and resulted from an earlier discovery by another Italian, Luigi Galvani (1737–1798). Galvani found that the leg muscles of a frog twitched when they were touched with a metal scalpel during a thunderstorm. He further showed that even in the absence of a thunderstorm the muscles twitched repeatedly if they were touched simultaneously with two dissimilar metals, such as copper and iron. This twitching, Galvani surmised, was due to a vital force called 'animal electricity'. However, Volta disagreed with Galvani's interpretations of the twitchings and proposed that the source of the electricity had nothing to do with a mysterious life-force, but that it was generated by contact between two dissimilar metals.

Volta demonstrated that two dissimilar metals in contact even with a simple salt solution were able to generate a continuous electric current. He placed a disc of cardboard soaked in a salt solution between a disc of silver and a disc of zinc and found that this simple device produced an electric current. Piles of such discs stacked one on top of the other produced even greater electric currents. Such 'voltaic piles' were the first ever electric batteries. Scientists could now easily produce relatively large amounts of electricity in a continuous flow, without resorting to storing up static electricity in spheres of sulphur.

In 1819, shortly before Faraday made his first prototype electric motor, the Dutch scientist, Hans Christian Oersted (1777–1851), discovered that if a wire connected to a battery is suspended horizontally over a compass needle, the needle moves whenever a current flows through the wire. This important phenomenon, which is called **electromagnetic induction**, clearly demonstrated that electricity and magnetism were related to each other. Very soon afterwards the German physicist, Johann Schweigger (1799–1857), invented the first galvanometer for measuring the current flowing through a wire.

It is in this scientific context that Michael Faraday began his studies of electromagnetism. However, in order to understand properly how Faraday made his discoveries, we need to appreciate not only the state of scientific knowledge when he began work, but also how he came to be a physicist and what motivated him to carry out his research.

Michael Faraday's background

Michael Faraday was born on 22 September 1791, in Newington, Surrey (now Elephant and Castle, London), the son of a poor blacksmith. The Faradays belonged to a religious sect called the Sandemanians (also known as Glasites). Sandemanians were non-conformist Christians who believed in adhering to a primitive Christian way of life. They considered wealth to be irreligious and God's laws to be written into Nature. By studying or 'reading' natural phenomena, they believed, humanity could discover God's true character. Michael Faraday remained a devout Sande-manian throughout his life and his approach to experimentation is consistent with the strong influence of his religion on his scientific studies. Indeed, the British physicist, John Tyndall (1820–1893), said, 'His religious feeling and his philosophy could not be kept apart; there was an habitual overflow of the one into the other.'

Faraday left school at the age of thirteen and started a job in London delivering newspapers for George Ribeau, a bookseller and bookbinder. Ribeau was impressed with the boy's work and in 1805 offered him a seven-year apprenticeship. Faraday accepted the apprenticeship, which was an opportunity for him to learn a new skill. More importantly, he came across many books on various subjects, some of which he read thoroughly and with great interest. A door to knowledge had opened up for the young Faraday. He was particularly inspired by articles on chem-istry. Even at this young age, and with little proper schooling, Faraday's interest in chemistry and physics was blossoming. As he said later, 'Whilst an apprentice, I loved to read the scientific books which were under my hands'.

During the apprenticeship, Faraday was able to carry out

simple scientific experiments in a spare room in the bookbinder's shop, with Ribeau's permission. Coupled with his bookbinding skills, the simple experiments he carried out provided Faraday with a good deal of manual dexterity that was later to be one of the hallmarks of his scientific research on electricity and magnetism. From 1810 onwards he attended lectures and discussions at the City Philosophical Society in London, which, together with his intense reading, gave him a firm background in basic chemistry and physics.

In 1812, Sir Humphry Davy (1778–1829), a British chemist and one of the world's most distinguished scientists of the time, gave a course of lectures at The Royal Institution in London. Faraday attended the lectures and was so enthused by them that he wrote them up, bound them into a volume, and sent them to Davy. With the volume, Faraday enclosed a letter asking Davy if there were any laboratory assistant posts available in the great chemist's laboratory. Davy was sufficiently impressed with Faraday's zest that he interviewed him, but offered him no job, recommending that Faraday stay in his post as a bookbinder. Not very long afterwards, however, Davy was temporarily blinded in a laboratory accident and he employed Faraday to take notes for him. In 1813, the 21-year-old Faraday gained his greatest break when Davy fired one of his laboratory assistants for quarrelling and offered Faraday a post as an assistant.

Michael Faraday, through his self-discipline, sense of responsibility and intense passion for knowledge, had entered the laboratory of one of the great scientists. He went on to become a very able scientist, at least as great as Davy. Indeed, it has been said that Faraday was Davy's greatest discovery, although the converse may also be said to be true, since they actually discovered each other.

Faraday still had enormous gaps in his education, having never been to university. In the same year that he was employed by Davy he set off with Davy on a tour of Europe, during which he met some of the finest chemists in the world, including Ampère and Volta. This tour, which lasted eighteen months, included daily tuition from Davy and probably more than compensated for Faraday's lack of university education. To have travelled with a great chemist and met and conversed with many others must have been the best exposure to scientific thought that Faraday could have had as a budding scientist.

Following his return from the European tour in 1815, Faraday assisted Davy and carried out experiments of his own. By 1820, Michael Faraday was an experienced experimentalist and a learned scientist with his own original way of looking at Nature. His route to this position had been unusual and unconventional. Some of the greatest scientific discoveries ever made were to follow, carving Faraday's name eternally into the history of human progress.

Faraday's discoveries

Oersted's discovery that a wire carrying an electric current causes a magnetic needle to move was a major step forward because it provided the first direct link between electricity and magnetism. Other scientists began to study this phenomenon. The French scientist, André-Marie Ampère (1775–1836), for example, demonstrated that two wires, each carrying an electric current, behaved as if they were two magnets: they could magnetically attract or repel each other. Sir Humphry Davy carried out experiments on electromagnetism with assistance from Faraday, whose curiosity about the subject was consequently stimulated.

In 1821, Faraday discovered a phenomenon called **electromagnetic rotation.** He devised an apparatus in which an electric current was able to cause a wire to rotate around a magnet and, conversely, a magnet was made to rotate around a wire. Figure 1 shows the basic design of this experiment. On the left-hand side, a magnet is pivoted so that it can freely rotate in a beaker of mercury. Mercury was chosen because it is a metal, and therefore conducts electricity; and, because it is a liquid, it allows any movement of the magnet. Dipping into the mercury is a fixed wire, which is connected to one end of a battery. The metal pivot to which the magnet is connected is attached to the other end of the battery. When a current flows through the wire it is conducted by the mercury so that a complete electric circuit is created. The vertically fixed wire becomes a magnet by electromagnetic induction. When the magnetic field around this wire interacts with the pivoted magnet, the magnet is made to move around the wire in a circular motion.

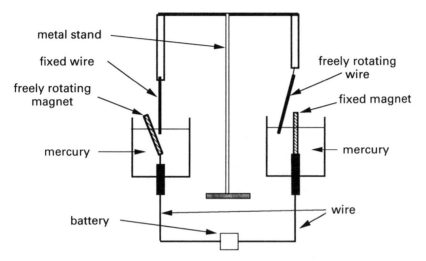

Figure 1.   Faraday's apparatus demonstrating electromagnetic rotation. On the left a pivoted magnet was made to rotate around a fixed wire when a current flowed through the apparatus; on the right, a freely suspended wire was made to rotate around a fixed magnet.

On the right-hand side (Figure 1), a magnet is fixed so that it cannot move. A vertically hanging wire, freely able to rotate, is now dipped into the mercury. When a current flows through the wire and mercury a magnetic field is induced around the wire. This field interacts with that of the fixed magnet and as a result the wire revolves around the magnet. Faraday interpreted the results of this experiment in terms of 'lines of force' that surround a magnet. According to Faraday, interactions between the lines of force of the magnets and those of the magnetic fields induced by a flow of electricity through the wires resulted in the rotations observed. The concept of these lines of force was later to be highly inspirational for James Clerk Maxwell's work on the mathematical analysis of the electromagnetic theory (Chapter 2).

This was the first time that anyone had ever produced continuous movement from an electric current. Faraday had converted the electrical energy produced by a battery into mechanical energy (movement of a wire or magnet). Any device that performs this transformation of electrical to mechanical energy is called an **electric motor.** Faraday's electromagnetic rotation apparatus was, indeed, the prototype electric motor, upon which much of our technology today depends.

It is astounding that this simple device should have such far reaching consequences and perhaps even more remarkable that Faraday made the first electric motor almost entirely because he wished to understand the laws of Nature that govern electromagnetism. His pursuit of physics had produced an unforseen benefit of gigantic proportions.

Simple though the apparatus for producing electromagnetic rotation may have been, many scientists failed to reproduce Faraday's experimental results, and more than a few scientists were sceptical of his findings. Faraday therefore made mini-models of the electromagnetic rotation apparatus and sent them, ready to connect to a battery, to some of his contemporaries. Those who received the models were left with no doubts that Faraday had, indeed, produced continuous electromagnetic rotation.

Faraday began to contemplate the possibility that magnetism might be able to generate electricity, since it was already known that electricity could generate a magnetic field. Following his discovery of electromagnetic rotation, he made several attempts to produce an electric current by placing a magnet in the vicinity of a wire. These experiments failed and it was almost another ten years before he successfully returned to them. Then, on 29 August 1831, Faraday made another device that was to have a great impact on humanity. He wound a coil of wire around one side of an iron ring and another, separate coil around the other side of the ring (Figure 2). Both coils were insulated so that there was no contact between them and the iron ring. One coil (*A* in Figure 2) was attached to a battery, the other coil (*B* in Figure 2) to a galvanometer. When Faraday ran a current from the battery through coil *A*, he made a remarkable observation. Immediately after the current had been switched on, the needle of the galvanometer attached to coil *B* moved in one direction, but it rapidly returned to zero even though current was still flowing through coil *A*. When the current was switched off, the galvanometer needle again momentarily moved, this time in the opposite direction to that which occurred when the current was switched on. The key to inducing a current in coil *B* was, Faraday realised, to *change* the current in coil *A*, rather than to maintain a constant current.

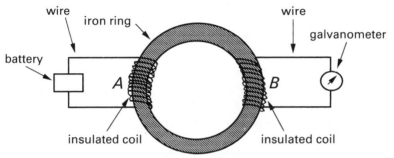

Figure 2. Faraday's transformer. A current was passed through a coil of insulated wire (*A*) that was wrapped around an iron ring. By repeatedly switching this current on and off, a current was induced in the secondary coil (*B*) of insulated wire.

Faraday clearly had induced a current in one coil by passing electricity through another coil. He correctly assumed that a current in coil *A* had induced a magnetic field around the iron ring, and that this magnetic field had induced an electric current in coil *B*. This was the first time that anyone had managed to convert magnetism into electricity, although the apparatus does involve an initial conversion of electricity to magnetism.

This device, again simple in design, was the first electric transformer. It allowed electricity to be generated in much greater amounts than had been previously possible: a small voltage in coil *A* could be made to produce a greater voltage in coil *B* if the number of turns in coil *B* was increased. In other words, low voltages could be transformed into larger voltages. Likewise, high voltages could be transformed into smaller ones. Transformers are used world-wide in thousands of electrical gadgets. They allow high voltages of electricity to be generated by power stations and these can then be converted to safer lower voltages for use in our homes. Every house, hospital, department store and office block depends upon transformers for its supply of electricity. And yet, Faraday made the first transformer as a result of his intrigue about the relationship between electricity and magnetism.

Following his work on the first transformer, Faraday carried out further experiments in which he managed to produce an electric current directly from a magnetic field, this time without using any source of electricity to initiate the phenomenon. He made several devices, one of which is shown in Figure 3. In this apparatus a magnet was moved in and out of a coil of wire. The movement caused a current to flow through the wire, as detected by a galvanometer attached to the coil. Again, the galvanometer needle returned to zero if the magnet was kept still. The more rapidly the magnet was moved in and out of the coil, the greater was the current induced in the wire, and the current flowed in opposite directions depending whether the magnet was moved in or out of the coil.

Faraday wrote of these studies, 'The various experiments prove, I think, most completely the production of electricity from ordinary magnetism.' He coined the term **magnetoelectric induction** for the generation of an electric current using a magnet.

Soon after these experiments, Faraday devised a machine in which a continuous electric current could be produced from a magnet. This device, shown in Figure 4, consisted of a large magnet with a copper disc between the magnet's poles. When the copper disc was rotated, a current was generated in the disc. This was the very first **dynamo**, a machine that converts mechanical energy (that which turns the copper disc) into electrical energy. In reverse form, when an electric current is used to turn a disc, the device is a direct ancestor of today's electric motor.

Magnetoelectric induction allowed, for the first time, the production of electricity without the use of a battery. In Faraday's time, batteries were expensive, heavy, and had to be replaced frequently. Magnetoelectric induction changed all this.

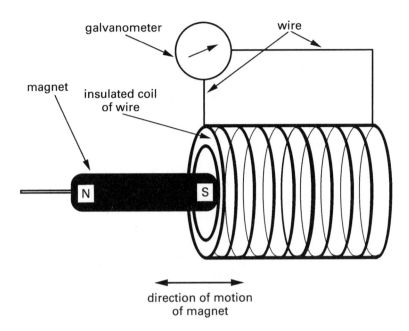

Figure 3.   Apparatus used by Faraday to demonstrate magnetoelectric induction. A magnet was moved in and out of a
coil of wire that was connected to a galvanometer: this motion
generated an electric current in the coil.

Faraday also made great contributions towards our understanding of matter. His studies of electrolysis, the process by which
liquid substances are broken down chemically by passing an electric current through them, led to his two Laws of Electrolysis.
These discoveries, which link atoms with electricity, were important for subsequent understanding of the structure of the atom
and of chemical reactions.

In addition to his work on electromagnetism and electrolysis,
Faraday made other important discoveries. He was the first
person to liquefy a so-called permanent gas (chlorine). It was,
until then, thought that this gas could never be liquefied. He
also discovered and isolated benzene, which was particularly
important for the dye and drug industries in later years, and he
determined the chemical composition of benzene. He was the first

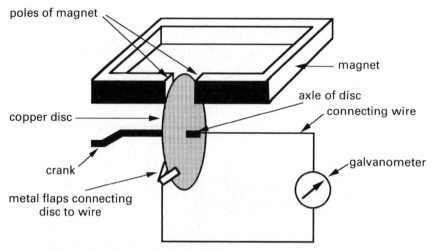

poles of magnet

magnet

axle of disc

connecting wire

copper disc

galvanometer

crank

metal flaps connecting
disc to wire

Figure 4. Faraday's dynamo, in which a copper disc was rotated manually between the poles of a powerful magnet, generating a continuous electric current.

scientist to make compounds of carbon with chlorine. Faraday's research on steel alloys helped lay the foundations of modern alloy research.

In 1845, when he was well into his fifties, he discovered the phenomenon of diamagnetism in which many substances, such as glass, which were thought to be non-magnetic, actually did show a small degree of magnetism when they were suspended near the poles of an electromagnet. The 'Faraday Effect', in which a magnetic field influences a beam of polarised light, was also one of his discoveries. Faraday's ideas about the physical nature of electromagnetism were an inspiration to the physicist, James Clerk Maxwell, who later developed the important electromagnetic theory of light (Chapter 2).

Several electrical units have been named in honour of Michael Faraday, including the 'farad', which is a measure of capacitance (the ability of a substance to hold electricity) and the 'faraday', a unit which measures the amount of electricity that causes chemical decomposition during electrolysis. Faraday's portrait appeared on the British twenty pound note in 1991; William Shakespeare previously held that honour. Prestigious awards for scientific merit and for the ability to communicate science are named after Faraday and he is considered to be the founder of the electrical engineers' profession.

Faraday's contributions to science and society are phenomenal. Yet he remained a dedicated and curious scientist who loved Nature and sought after Truth. He refused a knighthood and turned down offers of lucrative consultancies in order to pursue his research at a pure level. His desire, he said, was to remain 'plain Mr Faraday' to his death.

Michael Faraday was a genius who made enormous strides for physics as well as for society. However, he had no formal mathematical training, having left school when he was thirteen. Consequently, he saw electromagnetic phenomena in visual terms and made no attempts at analysing them mathematically. In the absence of a mathematical treatment, electromagnetic phenomena remained incompletely understood. It was not until 1864, thirty-three years after Faraday had made the first transformer, that the great mathematician, James Clerk Maxwell, finally interpreted electromagnetism in mathematical form. Maxwell's analysis led to the discovery of radio waves, X-rays and microwaves, and his ideas about the nature of light unexpectedly created another revolution in physics.

2
One giant leap for mankind

'That's one small step for a man, one giant leap for mankind.' These famous words, spoken by Neil Armstrong (b. 1930) on 20 July 1969, just as he became the first human being ever to step onto the Moon, were heard all over the world, even in people's living rooms. A century earlier, nobody could have realistically imagined that not only would a man walk on the Moon, but also that such a monumental achievement would be seen and heard by ordinary people as it happened. That the voice of Armstrong could be transmitted 400 000 kilometres (250 000 miles) through space into someone's house is so incredible that we may be excused for considering the feat something of a miracle. Yet it was the product of less than a century's worth of progress in radiocommunication, which began with the discovery of radio waves in 1888 by the German physicist, Heinrich Hertz (1857–1894).

Hertz's landmark discovery can be traced further back, particularly to a brilliant Scottish mathematician, James Clerk Maxwell (1831–1879). Maxwell predicted the existence of radio waves without even carrying out any experiments. His approach was purely theoretical, and it demonstrates the important influence that mathematics can have on human society and technological advancement. Napoleon Bonaparte (1769–1821), the first Emperor of France, said, 'The advance and perfection of mathematics are closely joined to the prosperity of the nation'. Maxwell's mathematical analyses of electricity and magnetism increased the well-being of many nations, and of the world as a whole, because it led to improvements in long distance communication on a scale few people could ever have imagined.

Maxwell not only forecast radio waves, but also gave a precise mathematical explanation for the electromagnetic phenomena that Faraday had so beautifully investigated. Maxwell opened up a new era of physics that paved the way for twentieth century developments such as quantum theory and Einstein's theory of relativity. He sowed the seeds for the discoveries of X-rays and microwaves, and for a host of other technological advances that now benefit humanity. How many of us realise that our microwave cookers had their origins in mathematical equations?

The early development of radiocommunication had three major phases. Maxwell's theoretical studies of electromagnetism dominated the first stage. The second phase involved the discovery of radio waves by Hertz, whilst in the third stage the technology of radiocommunication was improved and extended, especially as a result of pioneering work by the great Italian inventor, Guglielmo Marconi (1874–1937). We begin with James Clerk Maxwell and mathematics.

Maxwell and his equations

Maxwell was born in Edinburgh, Scotland, in 1831. As a youth he excelled in mathematics, presenting his first scientific paper, on geometry, when he was only fourteen. His background contrasts distinctly with that of Faraday: he was born into a wealthy family, had an excellent mathematics education and attended university.

Maxwell graduated from Cambridge University, England in 1854. He remained there as a graduate and began to study electromagnetism. He was particularly interested in Faraday's interpretation of electromagnetic phenomena. Faraday had suggested that the space around a magnet is filled with magnetic 'lines of force', which behaved rather like elastic bands. When these lines of force are vibrated, Faraday said, electricity is produced. Maxwell set himself the task of putting Faraday's discoveries and other known properties of electricity and magnetism into mathematical form. Maxwell was strongly inspired by Faraday's great book, *Experimental Researches in Chemistry*, in which electromagnetic phenomena had been described in detail. Maxwell later said, 'I shall regard it as the accomplishment of one of my principal aims to communicate to others the same delight which I have found myself in reading Faraday's "Researches".'

Maxwell, like Faraday, considered electricity and magnetism to be intimately related to each other. They should, he said, be envisaged to form an 'electromagnetic field'. In 1864, whilst he was a physics professor at King's College, London, he reached the climax of his mathematical analyses of electromagnetism. Using only his thinking capacity, his logic, and previous knowledge of the relationships between electricity and magnetism, he derived four mathematical equations. These are now called **Maxwell's Equations.** These four equations did more than describe electromagnetic phenomena accurately: they also created a revolution in physics.

One of Maxwell's main contentions was that an oscillating electric current should produce 'electromagnetic waves', and these waves should emanate from the electrical source and move through space. When he calculated the speed at which electromagnetic waves should travel, he was astonished to find that the value he obtained was virtually identical to that already determined for light waves. This speed, 300 000 kilometres per second, (186 000 miles per second), surely could not arise twice for two different phenomena unless the two were related? Maxwell correctly realised that, by using the power and beauty of mathematics, he had unified light with electricity and magnetism – something Faraday had strongly suspected decades earlier. 'Light itself,' Maxwell said, 'is an electromagnetic disturbance propagated . . . according to the laws of electromagnetism.'

The US mathematician, Richard P. Feynman (1918–1988), later summed up the enormous significance of this revelation, 'Maxwell could say, when he had finished with his discovery, "Let there be electricity and magnetism, and there is light!"'

Maxwell's electromagnetic wave theory not only predicted that light was an electromagnetic phenomenon, but also implied that hitherto undiscovered electromagnetic waves should exist. At that time visible light, ultraviolet light and infra-red rays were the only known forms of radiation. Maxwell's equations predicted that it should be possible to generate the 'new' electromagnetic waves in the laboratory from an electric current by oscillating the current repeatedly (changing the direction of flow of the current from one pole of a battery to the other and back again repetitively). The spectrum of electromagnetic radiation was, according to Maxwell, much wider than was known at the time.

The discovery of radio waves

Visible light was known well before Maxwell's time to consist of a spectrum of colours from red, through the colours of the rainbow, to violet. In 1801, the German–British astronomer, Sir William Herschel (1738–1822), discovered that light at the red end of the spectrum had a heating effect, then he found, much to his surprise, that even greater heat was produced beyond the red region of the spectrum. The radiation producing this heating effect could not be seen with the eye, and Herschel called the invisible waves **infra-red rays**. In the same year, the German physicist, Johann Wilhelm Ritter (1776–1810) discovered **ultraviolet radiation** beyond the violet end of the spectrum that was also invisible to the human eye. The spectrum of the early nineteenth century therefore ranged from infra-red, through the visible colours of the rainbow, to ultraviolet light. Things remained that way for more than three-quarters of a century, although Maxwell's wonderful equations had pointed the way for extension of what became known as the electromagnetic radiation spectrum: new types of invisible rays should exist on either side of infra-red and ultraviolet light.

Maxwell's equations predicted that the frequency at which the new electromagnetic waves should oscillate would be determined by the rate at which the electric current is oscillated. The higher the frequency of oscillation of the current, then the greater should be the frequency of vibration of the electromagnetic waves produced.

Many scientists did not take Maxwell's equations seriously and their revolutionary significance was not at first appreciated. Unfortunately, Maxwell died of cancer in 1879, nearly ten years before any of his predicted new types of electromagnetic rays were finally discovered. Verification of his ideas did come in 1888, when Heinrich Hertz produced and detected radio waves in the laboratory.

Hertz began to study Maxwell's electromagnetic theory in 1883. At some stage in his experimentation he was using a device called an induction coil, which produced an oscillating electric current. When this induction coil was attached to two conducting wires separated by a gap, a spark was produced across the gap. In 1888 he found that a spark produced in such a device actually caused another spark to appear across a second gap in a similar coil placed nearby (Figure 5). Further experiments confirmed this: if a second wire loop containing a gap but unattached to an electrical source was placed about one and a half metres (five feet) away from the first spark gap, a spark appeared in the second gap when the current associated with the first gap was switched on.

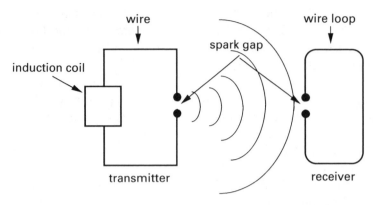

Figure 5. Equipment used by Hertz to produce and detect radio waves. An oscillating electric current produced from an induction coil caused sparks to appear across a gap in a wire attached to the induction coil. Radio waves were generated at this gap and were detected by their ability to induce sparks across a gap in a wire loop positioned some distance from the first gap. The induction coil and first spark gap constituted the prototype radio transmitter, whilst the second wire loop with its gap were the prototype radio receiver.

Hertz was convinced that this was evidence for the long predicted electromagnetic waves of Maxwell. He discovered that the signal coming from the first spark gap had the very properties predicted by Maxwell. It behaved as if it was electromagnetic radiation; for example, it travelled in straight lines and could be reflected from a metal sheet like light from a mirror. The waves he produced in this simple apparatus became known as radio waves.

Hertz had fulfilled Maxwell's predictions. His apparatus for producing and detecting radio waves was a prototype radio-communication system: the induction coil with its first spark gap was equivalent to a radio transmitter, whilst the secondary coil with its gap served as a radio receiver.

Even then, the possible applications of the discovery as a means of wireless communication were not clear. Hertz believed that the task of developing his apparatus as a means of long distance communication would meet with serious scientific problems. He was more happy with the fact that he had confirmed Maxwell's electromagnetic theory. In any case, Hertz died in 1894 at the young age of thirty-seven. The development of Hertz's apparatus into a genuine method of radiotelegraphy was due in large measure to Guglielmo Marconi.

Marconi and radiocommunication

Marconi's link with Hertz was Augusto Righi (1850–1920), a university lecturer at the University of Bologna in Italy, who was one of Marconi's teachers. Righi was involved in the improvement of Hertz's original transmitter; when Hertz died, Righi wrote an obituary in an Italian scientific journal. Marconi, who was only twenty years old at the time, read the obituary and became interested in Hertz's work.

Marconi was more of an inventor and engineer than an academic scientist. As such, he complemented Hertz and carried radio waves into their applied aspect. 'It seemed to me at this time that if this radiation could be increased, developed and controlled, it would most certainly be possible to signal across space, for very considerable distances', Marconi later said. Working in the attic rooms of his family's house, he elaborated upon Hertz's radio receiver and transmitter and managed to produce a secondary spark at a distance of 10 metres (30 feet), then 30 metres (100 feet). He increased the distance between transmitter and receiver to 3 kilometres (1.9 miles), and appeared to have no problems even if a hill was in between the transmitter and receiver.

At this point, Marconi contacted the Italian government for funding, believing that the enormous implications of his work would be appreciated. To his disappointment, he was refused financial support. Even other scientists found it difficult to believe that radio waves could be transmitted over long distances, particularly because they travel in straight lines: the curvature of the Earth should prevent the signals from ever reaching the receiver. Marconi refused to accept their reasoning and persevered with his endeavours to achieve radio signal transmission over very long distances. He turned successfully to Britain for financial backing of his project and moved there in 1896 in order to continue his work.

By 1897 Marconi was transmitting and receiving radio waves over a distance of 14.5 kilometres (9 miles) and, in 1898, he made the first radio transmission across the English Channel. In 1901, he flew a temporary receiving aerial on a kite in St John's, Newfoundland, and successfully received the letter 'S' in Morse code, which had been transmitted 3200 kilometres (2000 miles) away, from Poldhu, Cornwall, in England. This was the first successful transatlantic transmission of radio waves. Radio waves could indeed be transmitted over vast distances.

Rapid developments in the transmission of radio waves as a means of wireless communication followed. By 1906, refinements to the transmitting and receiving devices had been so great that it was possible for speech to be sent from one part of the world and heard anywhere else. Radiocommunication was particularly important initially for ship-to-shore communication. Electrical wires had been used to transmit messages on land from one place to another well before radio was discovered. Indeed, in 1876, Alexander Graham Bell (1847–1922), the Scottish–US inventor, patented the telephone which allowed sound to be transmitted as electrical impulses through wires. However, wireless communication was generally more practicable, particularly over long distances. In today's world, where radio is used in broadcasting (radio and television), industry, military and space communication and many more areas of everyday life, we may fail to realise that it began with James Clerk Maxwell's mathematical analysis of electromagnetism.

An interesting discovery that came about as an offshoot of Marconi's work on the radio was the identification of the **ionosphere**, a region of the Earth's upper atmosphere that contains charged particles. Without knowledge of the ionosphere, it was logical for scientists to be sceptical about the possibility of long distance transmission of radio waves from one country to another. Any radio waves should either miss the receiver because of the Earth's curvature or travel upwards and become lost in the upper atmosphere. In fact, radio waves are reflected off the ionosphere and travel back down to Earth: it is this reflection that allows them to travel such long distances successfully. Marconi had no really logical reason to believe that long distance radio transmission would work. But his long distance transmission was nevertheless successful, and his perseverance demonstrates that scientists should think carefully before they decide to give up an important experiment about which they may have some doubts. It is not an uncommon occurrence for an important discovery to be made even when the knowledge at the time predicts that the experiment will fail. In Marconi's case, the experiment worked because of the existence of the ionosphere, even though nobody was aware that it did exist.

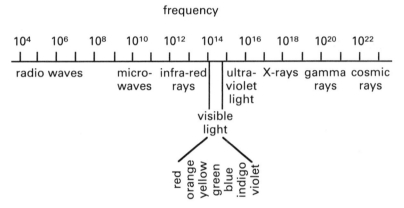

Figure 6. The electromagnetic spectrum showing the types of electromagnetic radiation. Before Hertz discovered radio waves, only infra-red rays, visible light and ultraviolet light were known. Maxwell's electromagnetic theory predicted the existence of other frequencies: radio waves were the first of these to be discovered. Frequency values are in oscillations per second.

Radio waves oscillate with frequencies much lower than those of visible light, infra-red and ultraviolet rays. The electromagnetic spectrum as it is considered to be today (Figure 6) covers a much wider range of frequencies than the spectrum known to Faraday and Maxwell. In addition to radio waves, radiation beyond the ultraviolet end of the spectrum was also predicted by Maxwell's equations. X-rays, gamma rays and cosmic rays were later discovered and further corroborated the electromagnetic wave theory.

Microwaves were also discovered between the frequencies of infra-red rays and radio waves and are now used in radar, telecommunication, medicine, astronomy and, of course, the microwave cooker.

Maxwell's equations are considered by today's scientists to mark the dawn of modern physics. By unifying light with electricity and magnetism, Maxwell changed the way in which physicists saw the world. According to Einstein, 'This change in the conception of Reality is the most profound and the most fruitful that physics has experienced since the time of Newton'.

Before the nineteenth century came to a close, X-rays were discovered, purely by chance, and when the twentieth century opened, Maxwell's theory was considered a triumph for physics.

3
Medicine's marvellous rays

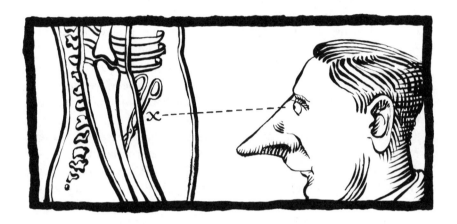

In addition to predicting the existence of radio waves, which vibrate at frequencies lower than visible light, Maxwell's equations also indicated that electromagnetic radiation with frequencies higher than ultraviolet light should exist. In 1895 a German scientist, Wilhelm Konrad Röntgen (1845–1923), identified such a new type of ray that startled the world and at the same time had a great impact upon medicine. His discovery of **Röntgen rays**, now more familiarly known as **X-rays**, is an excellent example of the important role played by luck in scientific research. It also illustrates Pasteur's dictum that 'chance favours only the prepared mind', because several physicists detected X-rays before Röntgen but failed to see their significance. Röntgen noticed something unusual, investigated it further, and realised he had discovered a novel and fundamental phenomenon.

Almost every hospital in every developed nation has an X-ray department and few people in these countries can say that they have never had an X-ray taken at some time or another in their lives. X-rays are used to detect bone fractures, accidentally swallowed objects, fragments of glass or shrapnel lodged in a wound, and even surgical instruments left inside a person during an operation. They can also detect cancerous tumours, and powerful beams of X-rays are used in the destruction of some cancers. The CAT (computerised axial tomography) scanner, a machine invented in the 1970s, is an X-ray device that allows tiny areas of tissues in a cross-section of the body to be examined using a computer. It is used today to identify diseased tissue with unprecedented precision. Countless lives have been saved, and medical problems prevented, as a result of Röntgen's chance discovery.

X-rays have also had an enormous impact on basic scientific research. For example, they are used to determine the three-dimensional molecular structures of proteins and other substances. Such studies are now being applied to the design of drugs for the treatment of a wide range of ailments from diabetes to heart disease, and it will not be too long before the fruits of this research are forthcoming. The wonders that X-rays have already performed for medicine represent the tip of the iceberg compared to what is in store for us. Industrially, X-rays are used to detect structural faults in machinery and buildings: they allow regions of intact objects to be visualised without having to take them apart. In the art world, X-rays have been used to examine paintings 'hidden' underneath subsequent ones made on top of the original. X-rays are used routinely to examine luggage being carried onto aircraft.

When Röntgen first described X-rays, this new and mysterious radiation was met with a great deal of trepidation and misunderstanding amongst the public and media. A London lingerie manufacturer advertised X-ray-proof underwear; a New Jersey politician introduced a bill to prevent the use of X-rays in theatre opera glasses; and there was a general apprehension that X-rays might penetrate the walls of homes and bring an end to a person's privacy. Such fears were unfounded and based on ignorance. The benefits of these new rays were forthcoming within a few months after their discovery, when a New Hampshire hospital used X-rays for the first time ever to diagnose a bone fracture. The rapidity with which X-rays were applied to the benefit of society is almost unparalleled amongst scientific discoveries: the fruits of pure research usually take much longer to appear. The scientific and medical implications of X-rays were tremendous, and Röntgen was rightly awarded the first ever Nobel Prize in Physics, in 1901, for his great achievement.

When he discovered X-rays, Röntgen was a fifty-year-old physics professor at Wurzburg University in Germany. Instrumental in this remarkable discovery was an apparatus called a **cathode ray tube**, or Crookes tube, which physicists had been using for some time to study the properties of electricity and matter. This device was developed as a result of earlier work carried out by Michael Faraday.

Cathode rays

Faraday had been interested in the possibility of detecting 'particles' of electricity. In 1838 he found that the negative pole (also called the **cathode**), but not the positive pole (**anode**), of a battery caused the region opposite to glow if the two poles were placed in a sealed glass tube and the air in the tube partially evacuated. Faraday suggested that the cathode was giving off an emanation that caused the glow. This apparatus was the forerunner of the cathode ray tube, numerous versions of which were devised by the British scientist, Sir William Crookes (1832–1919).

A cathode ray tube consists of a sealed glass tube with anode and cathode at opposite ends. An outlet from the glass vessel allows air or other gases to be introduced into, or evacuated from, the vessel (Figure 7). The apparatus is an ancestor of today's fluorescent light tubes and illuminated neon signs, and is a component of televisions. It is the essential constituent of the cathode ray oscilloscope, which is used to convert external signals (such as sound or movement) to an electric impulse and thence to a graphical representation on a screen. The cathode ray oscilloscope is one of the most extensively used tools in scientific research, industry and engineering.

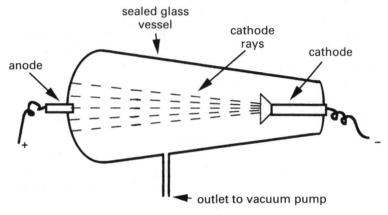

Figure 7. A cathode ray tube. The negative (cathode) and positive (anode) poles of an electricity supply were placed at opposite ends of a sealed glass vessel. When the electricity was switched on and the air evacuated from the vessel, a stream of electrons (cathode rays) was emitted by the cathode. These electrons impinged on the walls of the glass vessel, causing it to fluoresce. Röntgen discovered that X-rays were produced at the walls of the glass vessel at the point where cathode rays caused fluorescence.

It became clear towards the end of the nineteenth century that something was travelling from the cathode in straight lines and striking the anode opposite. When the anode was placed in a position that was not directly opposite the cathode, the emanations from the cathode missed the anode and instead struck the glass walls of the cathode ray tube directly opposite the cathode, creating a spot of luminescence on the glass. If an object was placed inside the tube in front of the cathode, a shadow of the object appeared on the walls of the vessel. These characteristics led some scientists to believe that the cathode was emitting a kind of radiation, and the emanations were therefore called **cathode rays**. Other scientists thought that the cathode rays were particles of electricity, particularly because they were deflected by the poles of a magnet, whereas ordinary light was known to be unaffected by magnets. We now know that cathode rays are beams of electrons, the negatively charged particles that carry electricity and are universal constituents of atoms.

Cathode rays were able to pass out of cathode ray tubes through small windows made of metal foil that were introduced into the wall of the glass vessel. They could be detected outside the tubes by their ability to cause fluorescence of certain chemicals placed a few centimetres from the metal foil windows. Fluorescence occurs when a substance emits light as a result of its being exposed to radiation. When the radiation is switched off the fluorescent substance ceases to glow. Using fluorescent screens, physicists showed that cathode rays were capable of penetrating ordinary air for only two or three centimetres (about an inch); beyond this distance they were lost, due to absorption by molecules in the air.

It was with this knowledge that Röntgen made his discovery of X-rays. He was particularly interested in the fluorescence produced by cathode rays and he also wanted to know if they could penetrate the glass walls of a cathode ray tube.

Röntgen's discovery

In November 1895, Röntgen covered the outside walls of a cathode ray tube with thin black cardboard in order to make sure that no stray fluorescence emerged from the walls of the glass vessel. The idea was to leave a small part of the glass wall uncovered and place a screen of fluorescent substance near this region: if cathode rays passed through glass, they should cause the fluorescent screen to glow. He set up the cathode ray tube, turned off the lights of his laboratory, and switched on the electricity supply to the tube.

There was certainly no obvious stray fluorescence from the walls of the vessel. However, through the corner of his eye, Röntgen saw a green glow. It was coming from an object more than a metre (three feet) away, too great a distance for cathode rays to have penetrated. He checked again for stray light coming from the glass walls and found no leakage: the green glow appeared again, and in the same place, when the electric current to the tube was switched on a second time.

At this point some scientists might have ignored the phenomenon and carried on with the particular experiment being conducted. But Röntgen had the insight to realise that he had stumbled upon something highly unusual and interesting. He knew that cathode rays could not possibly have been causing the green glow: they could not travel more than a few centimetres (one or two inches) in air. He investigated the source of the green light and realised that it was coming from one of his screens made of a fluorescent chemical. Something never described before was emanating from the cathode ray tube, penetrating over a metre (several feet) of air, and causing this fluorescent screen to glow.

Röntgen was so excited by this discovery that he spent the next six weeks working day and night in his laboratory, even sleeping there. During that short time he convinced himself that he had discovered a new type of radiation that was extremely penetrative. These rays, which he called 'X-rays' because of their unknown nature, were able to pass through paper (even a thousand-page book), tin foil, wood, rubber and many other substances, although lead seemed to stop them and some substances were more transparent than others. Röntgen also observed that 'If the hand is held between the discharge [cathode ray] tube and the [fluorescent] screen, the darker shadow of the bones is seen within the slightly dark shadow image of the hand itself.' This was the first time that anyone had seen the bones of an intact, living human hand. He subsequently showed that X-rays produce darkening of photographic plates, and he took the first true X-ray photograph, of his wife's hand, revealing her bones and wedding ring. In addition, he demonstrated that X-rays could be used to detect metal objects enclosed in wooden boxes.

Unlike cathode rays, X-rays were not deflected by a magnet and were capable of causing molecules in the air to become charged. However, although many of the properties of X-rays were comparable with those of light waves and other forms of electromagnetic radiation, Röntgen could not focus X-rays with a lens and he could not demonstrate their ability to diffract (bend round corners). It was not until more than ten years later that physicists managed to demonstrate properly the wave nature of X-rays and place them in their appropriate position in the electromagnetic spectrum (Figure 6). Because of the high frequency of oscillation of the waves and their short length (wavelength), X-rays will diffract only through extremely short gaps, such as those present between rows of atoms in crystals. Indeed, this diffraction of X-rays by crystals has allowed the molecular structures of many substances to be determined. **X-ray crystallography** is a powerful tool of modern chemistry and biochemistry: it has made a gigantic contribution to our understanding of enzymes, the proteins that carry out many of the biological processes of living cells and organisms. It also played a major role in determination of the structure of deoxyribonucleic acid (DNA), the molecule that holds the information that makes an organism what it is.

Röntgen made a chance observation of a green glow in the corner of his laboratory caused by radiation from his cathode ray tube impinging upon a fluorescent screen, and he recognised its significance. The discovery of X-rays is one of the many instances in which scientists have set out to study one phenomenon and have, during the course of their experiments, stumbled upon a new manifestation. **Serendipity**, the term used to describe chance discovery, was coined by the English politician, Horace Walpole (1717–1797). It comes from an old Persian fairy story, *The Three Princes of Serendip*, in which three princes regularly make fortuitous discoveries. (Serendip is an old term for Sri Lanka.)

In 1896, only months after Röntgen's discovery of X-rays, and as a consequence of it, another serendipitous breakthrough, the discovery of radioactivity, was made. Within a year, physics was transformed and humankind entered the atomic age.

4

Things that glow in the dark

In 1988 the Vatican allowed three laboratories in Switzerland, England and the USA to receive small fragments of the Shroud of Turin, a religious relic that was reputedly the burial cloth of Jesus Christ. Their aim was to determine the age of the shroud using a process called **carbon-14 dating** (also known as radiocarbon dating). The linen shroud, which had been kept in Turin, Italy, since 1578, had an apparently photographic image of the front and back of a bearded man imprinted upon it. Markings said to be due to the stigmata of Jesus as well as the crown of thorns and other lacerations and bruises also occurred on the image. Tests in the 1970s failed to show conclusively whether the markings on the shroud were painted with pigments or whether they were due to scorch marks or other processes. The powerful method of carbon-14 dating, which has been refined in recent years to give improved accuracy, should, it seemed clear, solve the riddle of the age of the cloth. Was the shroud's age the two thousand years required for it to carry the marks of Christ?

Remarkably, carbon-14 dating results from all three laboratories agreed that the Turin Shroud was made some time between AD 1260 and 1390. The Roman Catholic church accepted the data, and, thanks to carbon-14 dating, it is now known that the shroud cannot possibly have been associated with Jesus.

The case of the Turin Shroud demonstrates clearly how modern scientific procedures can solve archaeological problems. The carbon-14 dating method is based on the principle that living organisms accumulate small but measurable amounts of naturally radioactive carbon (carbon-14) in their tissues in addition to the large amounts of the common, non-radioactive form of carbon (carbon-12). The ultimate source of the carbon in all living things is carbon dioxide in the atmosphere. The ratio of carbon-12 to carbon-14 in atmospheric carbon dioxide remains essentially constant. Although each carbon-14 atom eventually decays and becomes non-radioactive, carbon-14 is constantly being replenished by nuclear reactions in which some of the nitrogen in air is converted into carbon-14 when it is bombarded by neutrons produced by cosmic rays.

When living organisms die, they no longer take up fresh atmospheric carbon dioxide. Therefore, the carbon-14 in a dead animal, plant or bacterium does become depleted over many years as it decays, whereas carbon-12, being non-radioactive, remains as it was when the organism died. In other words, the ratio of carbon-14 to carbon-12 falls as the length of time following the organism's death increases. The time taken for carbon-14 levels to fall by a half is known to be five thousand seven hundred and thirty years, so that the ratio of carbon-14 to carbon-12 will fall to half of its original (point of death) value in five thousand seven hundred and thirty years. By measuring the ratio of carbon-14 to carbon-12 in a dead organism, the year in which the organism died can therefore be estimated. Carbon-14 dating is widely used in archaeology and geology. Any once-living matter, including wood, charcoal, cloth (such as the linen of the Turin Shroud), bones, shells, animal and plant tissues, can be processed by the technique. Dates can be established for ancient human settlements, Egyptian tombs, geological rock strata and many more items of historical, archaeological and geological value.

Carbon-14 dating is one of many applications of the knowledge of radioactivity that has accrued in the twentieth century. Basic biological, medical and chemical research has gained tremendously from this information, and practical medicine has also greatly benefited. Treatment of cancer by destroying malignant tumours with the rays given off by radioactive substances came soon after the discovery of radioactivity. Radium was initially used for cancer radiotherapy, but other, artificial radioactive chemicals such as cobalt-60, are used nowadays. Radioactive substances may also be injected into the body and allowed to reach a particular organ, where their radiation can be detected. Such methods have been used to diagnose many disorders, including brain tumours, thyroid problems and heart disease.

Radioactivity has, indeed, come a long way since it was discovered in 1896 by the French physicist, Henri A. Becquerel (1852–1908). As with Röntgen's discovery of X-rays, radioactivity demonstrates wonderfully how pure science and seemingly simple discoveries can develop a phenomenal diversity of applications over a period of many years. The discovery of radioactivity was also serendipitous, and it again demonstrates that chance discoveries do require keen observation by the scientists involved.

The story of radioactivity is one of the most romantic developments in the history of science. Two physicists, Marie Curie (1867–1934) and Pierre Curie (1859–1906), a husband-and-wife team, followed up Becquerel's discovery with enthusiasm, perseverance, passion and personal hardship. Marie Curie was the first truly great woman scientist the world had seen, and she made the female entrance into the fore of science with style, receiving not one, but two, Nobel Prizes for her famous work. She inspired many future female scientists to break the barriers of male-dominated science.

The impact of the discoveries of Becquerel and the Curies on science and society was breathtaking. Apart from the plethora of applications to basic research, medicine and industry, nuclear fuel emerged with its potential to supply almost limitless energy to society, but also with its dangers; and the atom bomb was in its germination phase. The story of radioactivity begins in earnest with Henri Becquerel.

Becquerel's discovery

Henri Becquerel was appointed as a professor of physics at the Ecole Polytechnique in Paris in 1895. Soon after his arrival there he discovered radioactivity. He had been interested in the phenomena of fluorescence and phosphorescence, that is, the ability of some substances to glow when they are irradiated with visible light or other electromagnetic radiation. Whereas fluorescent materials glow only whilst they are being irradiated, phosphorescent objects glow for some time after irradiation. Röntgen had observed that X-rays emanated from the glass walls of cathode ray tubes, which fluoresced when they were bombarded with cathode rays (Chapter 3). Becquerel wondered if fluorescent or phosphorescent substances generally might emit X-rays, in which case it would be possible to produce X-rays without using cathode ray tubes.

In order to test this hypothesis Becquerel chose certain uranium salts that phosphoresced when they were exposed to bright sunlight. If his hypothesis was right, he should be able to detect X-rays coming from the uranium salts after they had been placed in the sun for several hours.

Becquerel wrapped a photographic plate in black paper to prevent any light from exposing it, then he placed crystals of a uranium salt on top of this package. The salt was exposed to the sunlight for several hours to allow it to phosphoresce and the photographic film was then developed. Becquerel found a clear image of the salt on the film, indicating that it had given off radiation that had penetrated the paper and blackened the photographic emulsion. This result clearly agreed with his hypothesis that the phosphorescent uranium salt was emitting X-rays, since

X-rays had been found by Röntgen to have this very effect on photographic film through a layer of paper.

Becquerel repeated the experiment in order to confirm his exciting finding. However, a series of fateful events occurred in which the weather conditions in Paris changed the course of history for the better. Becquerel again wrapped a photographic film in black paper and placed the uranium salt on top of the paper, but when he came to expose the salt to sunlight he found that the clouds had covered the Sun. He waited for the Sun to shine again, but it failed to do so on that day. Eventually, Becquerel decided to place the salt and photographic film in one of his laboratory drawers until the Sun did shine, when he would continue with the experiment.

But the Sun did not shine in Paris for another three days and Becquerel, eager to continue his experiments, decided anyway to develop the photographic film. He expected to see a faint image of the uranium salt on the film, since it had been exposed to weak light and might have phosphoresced to a small degree. To his great surprise, however, he found that the salts had produced a very intense blackening of the photographic emulsion, far greater than was produced even by exposure of the crystals to very strong sunlight for several hours. He realised that the uranium salt might have emitted rays irrespective of its phosphorescence. He confirmed this by showing that the salt caused blackening of photographic emulsion even when it was placed on top of film in the dark without any prior exposure to light at all.

Light was not needed for the production of these rays: the uranium salts were emitting them spontaneously. Becquerel wrote, 'The radiations are emitted not only when the salts are exposed to light, but also when they are kept in the dark, and for two months the same salts continue to emit, without noticeable decrease in amount, these new radiations.'

Becquerel went on to demonstrate that spontaneous emission of radiation was a property of all uranium salts and, by showing that pure uranium was a more powerful emitter than its salts, he concluded that uranium was the source of the rays, and not other components of its salts. (We now know that the radioactivity of pure uranium increases with time because the uranium decays to another, more radioactive element: this, rather than the uranium itself, is the major source of the radioactivity.) He found that the rays did resemble X-rays in their penetration power, their ability to darken photographic emulsion and their ability to induce molecules in the air to become charged. It was later shown that several types of emissions were responsible for the effects observed by Becquerel, and gamma rays, a type of electromagnetic radiation with a frequency greater than that of X-rays, were responsible for the X-ray-like properties of Becquerel's rays.

This was an unusual phenomenon to physicists – it was something that they found difficult to explain. Never before had anyone shown that radiation could spontaneously emanate from matter. Becquerel had also shown that the radiation given off by uranium was not affected by the chemical state of the uranium or by physical conditions such as temperature. The emanation of radiation seemed to be completely unaffected by changes in the surrounding environment.

Few scientists had any idea of the tremendous implications of Becquerel's discovery. However, another physicist, Marie Curie, who was studying in Paris, did have the insight necessary to appreciate its significance. She coined the word, 'radioactivity' for the phenomenon and, with her husband, the French physicist, Pierre Curie, discovered and isolated two new radioactive elements. The 1903 Nobel Prize for Physics was shared between Becquerel, 'for his discovery of spontaneous radioactivity' and the two Curies, 'for their joint researches on the radiation phenomena discovered by Professor Henri Becquerel'.

Marie and Pierre Curie

Marie Curie was born Marya Sklodowska in Warsaw, Poland in 1867. She left to study physics in Paris in 1891 and gained the top physics degree of her year at the Sorbonne. She had planned to return to Poland to teach, but instead married Pierre Curie, a physicist who was also working in Paris. It was decided that Marie would do her doctoral research, without pay, in Pierre's laboratory. The Curies were interested in Becquerel's new discovery, and Marie and Pierre decided that her doctoral thesis should be on the nature and origin of Becquerel's rays.

Marie Curie believed that the emission of radiation by uranium was a property of uranium atoms that had nothing to do with its chemical reactivity. This idea that radioactivity was an 'atomic' property that might reveal something fundamental about the internal structure of atoms was one of great insight, and it was a major reason why Marie Curie chose radioactivity as her research topic. Here was her chance to work in virgin territory in an area that promised exciting revelations about the fundamental properties of matter.

Within only a few weeks of starting her doctoral research, which she carried out in a poorly equipped, cold and damp laboratory that had been previously used as a storeroom, Marie showed that the levels of radioactivity were related to the amount of uranium in her radioactive samples, and she verified Becquerel's work showing that the amount of radiation given off was not affected by the chemical or physical state of the uranium.

Marie searched for other substances that might be radioactive besides uranium. By screening samples of many chemicals and minerals containing the elements that were known at the time, she discovered that another element, thorium, was also radioactive. Uranium was not alone: radioactivity was more widespread than at first was realised.

When she examined more minerals, particularly samples of pitchblende (a uranium ore), she found that the amount of radioactivity in some of them was far greater than could be accounted for by their content of uranium and thorium, and she believed that there must be another radioactive element or elements in these minerals. Because she had examined minerals containing all known elements, she concluded that these additional radioactive substances must be new, as yet undiscovered elements.

Other physicists thought that she had made a mistake, but Pierre agreed with her. Both of the Curies believed they had evidence for the existence of at least one new element. 'The physicists we have spoken to believe we have made an error in experiment and advise us to be careful. But I am convinced that

I am not mistaken', Marie wrote. This advice from other physicists could not have been given to anyone more meticulous and careful than the great Marie Curie, who even wrote down in her log book the exact time of day when she carried out experiments and often recorded the temperature of the cold and dismal laboratory in which she worked. With the experience and insight of Pierre complementing her, how could there be any doubt as to the validity of their results?

In 1898 Marie and Pierre Curie announced the probable presence of a new radioactive element in pitchblende ores, and they set about isolating it. For the rest of her doctoral research, Marie did her laboratory work alongside Pierre, who realised its importance. By breaking down pitchblende ore into its constituent chemicals, they found that there was not one, but two, new elements that were responsible for the high levels of radioactivity in the ore. The first of these new elements to be named was **polonium**, in honour of Marie's beloved Poland. The second element, which they later called **radium**, was even more radioactive than polonium and its isolation took four long and arduous years.

Radium was present in extremely low quantities in pitchblende (less than one part in a million by weight). In order to purify it from the ore, enormous amounts of pitchblende were needed. The Curies were already suffering financial difficulties, having themselves and a child to support on Pierre's meagre salary. Pitchblende was the commercial source of uranium, but it was expensive and Pierre was unable to obtain the necessary funds from his research institute. However, Marie realised that they could obtain both radium and polonium from the residue of pitchblende from which uranium had been industrially extracted. This residue was discarded by industry and was available cheaply: it was the ideal source of the new elements.

Marie and Pierre paid for the transport of a ton of pitchblende residue from mines in Bohemia to their laboratory in Paris. They asked for a bigger laboratory in which the purification of polonium and radium could be carried out, but were refused it. Instead, they were allowed use of an abandoned shed near to their existing laboratory. This shed, which had a leaking roof and a floor consisting of flattened earth, and which lacked any proper scientific equipment, was the site of one of the most heroic scientific endeavours in history. There can be few laboratories so poorly equipped and under-resourced that have produced Nobel Prize-winning research.

The Curies lived a life of financial hardship and worked under appalling conditions, yet by 1902 they had isolated one-tenth of a gram of pure radium chloride from a ton of pitchblende. Marie and Pierre proved beyond reasonable doubt that radium was, indeed, a new element. Any sceptical physicists remaining were finally satisfied that the Curies had made a great discovery.

During the period in which they isolated radium, Marie and Pierre would occasionally go down to their shed during the night to observe the bluish glow of the radium they had discovered. Despite the harshness of the times, Marie later said, 'It was in this miserable and old shed that the best and happiest days of our life were spent, actively consecrated to work. I sometimes passed the whole day stirring a boiling mass, with an iron rod nearly as big as myself. In the evening I was broken with fatigue.'

In 1903 Marie presented her doctoral thesis. Soon afterwards, Marie and Pierre Curie received the Nobel Prize for Physics. Although the work on radioactivity carried out by the Curies was a joint venture and credit must go to both of them, Marie's achievements were particularly remarkable. Not only was she a female scientist – a rarity in those days – but also she obtained a Nobel Prize from her doctoral work! She was the first woman to receive a Nobel Prize, and eight years later was to receive a second Nobel Prize, this time in Chemistry, for her research on the isolation and properties of polonium and radium. She received the second Nobel Prize without her husband, who tragically died in 1906 during a road accident in which one of the back wheels of a horse-drawn carriage crushed his skull, killing him instantly.

On numerous occasions, Pierre Curie applied for a professorship and for funding for a properly equipped laboratory, but his pleas were always turned down. On a visit to England, when the Curies were already well known as distinguished scientists, Pierre attended a lavish dinner and noticed the expensive jewellery worn by some of the wealthy ladies there. He reputedly spent a good proportion of the evening calculating how many laboratories he could purchase for the price of those necklaces and rings.

Marie and Pierre Curie were not only brilliant and dedicated scientists, but they were also human beings with the highest moral standards. They could easily have made a fortune by patenting their method for isolating radium, which fast became an industry, but instead they freely gave the patentable information to any person or company who requested it. Marie said, 'If our discovery has a commercial future, that is an accident by which we must not profit. And radium is going to be used in treating disease . . . It seems impossible to take advantage of that.' Pierre agreed, 'It would be contrary to the scientific spirit.' They also shunned fame, to the extent that they did not even attend the Nobel Prize ceremony in 1903: instead they sent a letter to the organisers in Stockholm saying that they had too much teaching to do. They postponed their Nobel lecture until the following year.

Despite their difficulties, the Curies made do with the dilapidated conditions available for their research and their perseverance paid off in the end. Now that he was a Nobel laureate, Pierre was finally appointed, in 1904, as a professor of physics at the Sorbonne and he was offered the chance to fulfil his lifelong wish for a well equipped laboratory. Sadly, he never lived to see the new laboratory, which was not completed until after his death. Marie took over his professorship and so became the first woman ever to be a professor at the Sorbonne.

Many scientists who worked with radium, including Marie and Pierre Curie, began to realise that it was a potentially dangerous substance to handle without taking appropriate precautions. Marie Curie did so much work with large amounts of radium that she exposed herself to enormous doses of radioactivity during her career. Eventually, radiation sickness, later to be so tragically evident in victims of the atomic explosions in Hiroshima and Nagasaki (August 1945) and of the Chernobyl nuclear accident (April 1986), also took its grip on Marie's health. As a result of this radiation exposure, she contracted leukaemia and, in 1934, she died. Ironically, one of the benefits of radioactivity was its use in the treatment of cancerous tumours, including leukaemia.

A noteworthy step forward in the study of radioactivity was

made in 1933 by Pierre and Marie Curie's daughter, Irène Joliot-Curie (1897–1956), in collaboration with her husband, the French physicist, Frederic Joliot (1900–1958). They discovered **artificial radioactivity**, demonstrating that non-radioactive elements could be converted to radioactive ones in the laboratory. This opened the way to production of hundreds of new radioactive substances that have been of great value to science, medicine and industry. For example, iodine-131, an artificial radioactive substance, is used to diagnose diseases of the thyroid gland, and pernicious anaemia can be diagnosed by labelling Vitamin B12 with cobalt-60 (cobalt is part of the molecular structure of Vitamin B12) and examining the excretion of the cobalt-60 in the urine. Irène, like her mother, received the Nobel Prize for Physics, which she shared with Frederic Joliot in 1935, and she, too, died of leukaemia as a result of overexposure to radioactivity in the laboratory.

Radioactive decay

Marie Curie was correct when she had the idea that radioactivity was an atomic property and that it might be a source of important information about the structure of the atom. Following the discovery of polonium and radium, it eventually became apparent that radioactivity was a property of the atom's nucleus, rather than being related to the electrons around the nucleus. At that time, it was appreciated that electrons were involved in chemical reactions between substances, but little was known about the other components of the atom or about their arrangement within the atom. Later studies demonstrated that the nucleus occupies a minuscule proportion, only a hundred-million-millionth, of the volume of an atom, even though almost all of the atom's mass is present in the nucleus.

Radioactivity occurs when an atom's nucleus is unstable: the nucleus undergoes radioactive decay into a more stable one. Since the identity of an element is determined by its nucleus, radioactivity is a process by which one element is converted to another.

The nucleus of a uranium atom, for example, is unstable and so uranium decays into thorium. This thorium also has an unstable nucleus and decays into another element, protactinium. The process of transmutation of elements from those with less stable nuclei to those with more stable ones continues by radioactive decay until a non-radioactive (stable) element is produced. Uranium can decay into a non-radioactive form of lead in fourteen steps; radium and polonium are intermediates along this radioactive decay series.

When nuclei undergo radioactive decay they can emit various types of radiation or particles. Three major types of emissions, and those recognised in the early part of the twentieth century as emanating from uranium and its daughter elements, are **alpha particles, beta particles,** and **gamma rays**. Alpha particles are helium nuclei and are relatively heavy; they represent small chunks of the heavier radioactive nuclei. Beta particles are electrons that originate in the nucleus, whilst gamma rays are electromagnetic radiation of high energy. Radioactivity provided a clear way of studying atomic nuclei: examination of the emanations from unstable nuclei led the way to a more complete understanding of the structure of matter.

5

Parcels of light

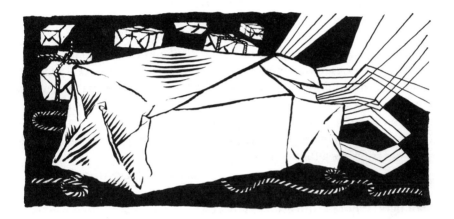

Radioactivity came as a surprise to most physicists. The emission of a seemingly endless flow of radiation energy from radioactive atoms appeared to be at odds with the widely accepted law that energy cannot be created or destroyed. Where was all of the energy coming from that was being given off by radioactive substances? (The solution to this enigma eventually came from Albert Einstein (1879–1955), who showed mass could be converted to energy (Chapter 6). During radioactive decay, a tiny amount of matter in the nucleus of an atom is converted to a very large amount of energy, which is emitted as radiation.)

Despite this puzzle of radioactivity, many physicists living in the late nineteenth century believed that they had reached an almost complete understanding of the physical world. Maxwell's electromagnetic theory, in particular, was a triumph for physics because it brought together light and electromagnetism and explained many phenomena; it also predicted the existence of new types of rays, which were subsequently found in the form of radio waves and X-rays. Maxwell's theory implied that light consisted of continuous waves of electromagnetic disturbances, and few, if any, physicists doubted this idea. It explained most of the properties of light, especially the phenomena of **diffraction** (ability to bend round corners) and **interference** (the ability of light to split into two portions which then meet each other and enhance or cancel each other out). The theory of Sir Isaac Newton (1642–1727) that light was made up of minute particles ('corpuscles') seemed to be well and truly dead and buried. Light was made up of electromagnetic waves and not particles, everyone agreed. In addition to electromagnetic radiation, there was matter, which was made up of atoms; and there was gravity, which was somewhat less understood. Newton had described mathematically how objects move and he had produced a mathematical equation to describe the effects of gravity.

Nevertheless, there were several features of electromagnetic rays that were not clearly explicable in terms of the wave theory of light. One was the precise way in which light was emitted when objects were heated to very high temperatures. The other was the observation that light causes metals to emit electricity. When these phenomena were studied in detail they were found to conflict with the idea that light is made of waves, and they led to the re-emergence of the particle theory of light. This constituted a transition between the so-called **classical physics** of the eighteenth and nineteenth centuries and the **quantum physics** of the twentieth century.

The founder of quantum physics was a German theoretical physicist called Max Planck (1858–1947), who was a professor at Berlin University. Revolutionary though his great theory was, it took five years before anyone took any notice of it. Then, one of the greatest scientists of all time, Albert Einstein, revealed the importance of Planck's ideas. Physics and our understanding of the Cosmos in which we live have not been the same since.

In order to understand why the particle theory of light was revived, it is necessary to have some knowledge of the two phenomena, the photoelectric effect and blackbody radiation, which were responsible for the re-emergence of the idea.

The photoelectric effect

In 1887, whilst Hertz was carrying out the experiments that led him to discover Maxwell's predicted radio waves, and so provide apparent confirmation of the electromagnetic wave theory of light (Chapter 2), he also discovered another phenomenon that was later found to contradict Maxwell's theory. Hertz found that light given off by the first spark gap in his prototype radio transmitter and receiver apparatus (Figure 5, Chapter 2) enhanced the spark that was induced in the secondary (receiver) spark gap. Indeed, when he shone a lamp directly onto the secondary gap, the sparks were even stronger. Somehow, light was having an effect on the electric current flowing in the secondary spark gap.

A year later, the German physicist, Wilhelm Hallwachs (1859–1922), discovered that ultraviolet light caused a negatively charged zinc plate, but not a positively charged one, to become neutral. The explanation for this result, and for Hertz's observation, became clear when the British physicist, Sir John Thomson (1856–1940), and the German physicist, Phillipp Lenard (1862–1947), demonstrated that light causes the ejection of electrons from metal surfaces. In Hertz's experiments, light was causing electrons to be emitted from the metal poles of the second spark gap, producing sparks in addition to those produced by the radio waves. In Hallwachs' experiments, electrons (which are negatively charged) were ejected by light from the negative zinc plate, neutralising its charge. The process by which light caused emission of electrons from metal surfaces became known as the **photoelectric effect**.

The photoelectric effect not only played a central role in theoretical physics during the early part of the twentieth century, but also produced many applications. The television, photographic light meters, burglar alarms, automatic doors, electronic switches and solar-powered equipment are some examples of devices that exploit the conversion of light to electricity.

According to Maxwell's wave theory of light, one would predict that if the intensity of light incident upon the metal surface was increased, electrons with a greater speed (higher energy) should be emitted from the surface. Changing the frequency of light should have no effect on the energy of electrons ejected, provided that the light intensity remained the same. In order to check this prediction, Phillipp Lenard examined the effect of altering the intensity and frequency of incident light on the release of electrons during the photoelectric effect.

Lenard was extremely surprised to find that, contrary to what was expected from Maxwell's theory of the wave nature of light, increasing the intensity of light actually had little effect on the energy of electrons ejected. Instead, as the intensity of light was raised the *number*, rather than the energy, of electrons emitted from the metal surface was increased. The frequency of the light did, however, affect the electrons' energy. At high frequencies the energies of all electrons emitted were equal to each other; lowering the frequency but maintaining the same light intensity resulted in the same number of electrons being emitted, but their energy was lower.

This peculiar result, which appeared to be in direct conflict with the widely accepted idea that light travels through space as continuous waves, remained a puzzle for many years. When an explanation did come for the photoelectric effect, the two theoretical physicists, Max Planck and Albert Einstein, provided the answer to the puzzle, and a new era of physics was born.

Blackbody radiation

Another set of experimental results that appeared to contradict the nineteenth century physicists' ideas regarding light as a wave travelling through space involved studies of **blackbody radiation**. A blackbody is an object that absorbs all of the electromagnetic radiation, regardless of its frequency, that strikes it, and all of this radiation is emitted again when the blackbody is heated to high temperatures. Most people are aware that when an object is heated it not only gets hotter, but also its colour changes from red through orange and yellow to white. Stars also glow with different colours, from red through yellow to white, then blue, as their temperatures increase. This is the same as saying that the frequency of light emitted increases as an object is heated. Knowledge of blackbodies has allowed the temperatures of stars, including our own Sun, to be determined from the frequencies of light they emit. The intensities and frequencies of light emitted by the Sun, for example, are identical to those expected to be emitted by a blackbody with a temperature of 6000 degrees Centigrade, indicating that this is the temperature of the surface of our Sun. Knowledge of blackbodies has also been influential in interpreting evidence for the origin of the Universe in the Big Bang (Chapter 7): the early Universe behaved as a blackbody and the cosmic background microwave radiation has the frequency and temperature predicted for blackbody radiation that would have been left over from the early Universe.

True blackbodies were difficult to find but a close approximation was suggested by the German physicist, Wilhelm Wien (1864–1928). Wien proposed that a lightproof container with a small hole in its wall would behave as if it were a blackbody (Figure 8). If light is shone through the hole in such a container, all of the light will eventually be absorbed by the inner walls. The chances of the light leaving by the hole are very small, so even if the light is initially reflected it will strike other parts of the wall until it is finally absorbed. In other words all light entering the hole, regardless of its frequency and intensity, will be absorbed – exactly the situation required of a blackbody. When the container is then heated, this absorbed light will be given off from the walls and the radiation that is emitted through the hole can be examined as blackbody radiation.

In the middle of the nineteenth century the German physicist, Gustav Kirchhoff (1824–1887) had demonstrated that particular substances absorb light of a specific frequency. Sodium, for example, absorbs yellow light, whilst potassium absorbs violet light. He also found that when these substances are heated they emit light of the same frequency as that which they absorb. So sodium gives off yellow light when it is heated and potassium emits violet light. A blackbody, therefore, should emit light of all frequencies when it is heated.

Wien was particularly interested in the precise way the energy

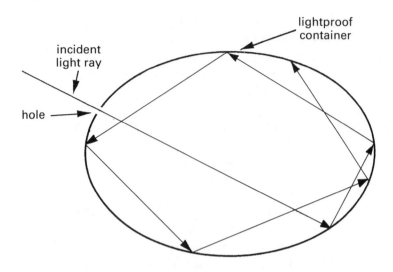

Figure 8.   Model used as a blackbody. Wilhelm Wien proposed that a container with a small hole in its walls would
simulate a blackbody. Any light entering the container through
the hole is absorbed by the inner walls. If the light is initially
reflected from the wall, it will eventually strike another part of
the wall and be absorbed, since the chances of its leaving via
the hole are negligibly small. If the container is then heated,
blackbody radiation is emitted from the hole and its frequency
and energy can be measured at different temperatures.

and frequency of light was emitted when a blackbody is heated.
He found that at a given temperature the electromagnetic radiation emitted from a blackbody was more likely to be of one
particular frequency than any other frequency. When the temperature of the blackbody was increased, more light energy was
emitted at all frequencies, but again one particular frequency
dominated, and this was higher than the frequency that dominated at the lower temperature.

Wien developed a mathematical formula to explain his results.
However, whilst it did explain the results that he obtained with
the high frequencies of light emitted from a blackbody at various
temperatures, it failed to provide an explanation for the results he
obtained with the low frequency radiation. Another mathematical

equation was already available that did account for the low frequency data, but this formula failed to explain the high frequency results.

In other words, there were two mathematical equations, each of which only half-explained the experimental data. What was needed was some way of merging these formulae to produce an equation that accounted for all of the data obtained for both high and low frequency radiation.

Max Planck and quantum physics

Max Planck was intrigued by blackbody radiation. He had the great insight to formulate the required mathematical equation from the two equations that only half-explained the data. Planck did not derive the equation theoretically or experimentally: he simply 'guessed' it. That is not to deny the brilliance of Planck: this was an inspired guess that required a mind of great genius.

Nevertheless, Planck knew that he had to derive the new equation theoretically from first principles in order to convince other scientists of its validity: simply guessing it was not enough. He therefore set about the task. When he did this, he came to the realisation that the only way he could logically end up with the right equation was by assuming that light was taken up and given off by a blackbody not as a continuous wave, but in little 'parcels'. He called these parcels of light energy, '**quanta**' (single, '**quantum**'). In 1900, quantum physics was born.

According to Planck, blackbodies absorb and emit light in small packets called quanta. Einstein later likened Planck's revelation to a beer barrel that produced beer only in pint portions rather than in a continuous flow. Here, the tap of the beer barrel was equivalent to the blackbody, the beer to the radiation emitted and absorbed, and the pint portions were analogous to the quanta: the tap allowed only pint portions to flow from it.

Planck's theory was seen to provide an explanation for blackbody radiation, but it was largely neglected by physicists for another five years. It is one of those cases of a monumental scientific breakthrough that was way ahead of its time being ignored by scientists caught up with the more established theories of their time. Then, in 1905, quantum physics soared to the fore of physics as a result of the work of a theoretical physicist who did not even hold a university post, but who worked as a clerk in the Swiss Patents Office in Berne, Switzerland. This physicist, who was possibly the greatest scientist the planet has ever seen, created a revolution in physics at least as significant as that produced two centuries earlier by Sir Isaac Newton. His name was Albert Einstein.

Einstein and the photon

In 1905, Albert Einstein published four theoretical papers in distinguished scientific journals. Two of them were concerned with his Theory of Relativity (Chapter 6). Another was on **Brownian motion**, which refers to the random jerky movements of microscopic particles suspended in air or a liquid. The fourth paper proposed an explanation for the photoelectric effect, and it was this article that revived the long-forsaken particle theory of light and brought Planck's ideas into the scientific limelight. It has been said that any one of the four of Einstein's 1905 papers would have made him an eminent physicist. Most people remember him for Relativity, although when he later received the Nobel Prize for Physics, in 1921, it was for his theoretical work on the photoelectric effect.

Einstein was well aware that the wave theory of light did not explain the photoelectric effect. 'The usual idea that the energy of light is continuously distributed over the space through which it travels meets with especially great difficulties when one tries to explain photoelectric phenomena', he said. He was acquainted with Planck's explanation of blackbody radiation, that light was emitted and absorbed as quanta of energy, and it was to this that Einstein turned in order to solve the photoelectric effect problem.

The solution, Einstein proposed, was to regard light as being made up of particles or quanta in much the same way that Planck considered light to be absorbed and emitted from blackbodies as quanta. These quanta of light later became known as **photons**, from the Greek word meaning 'light'. However, whilst Planck believed that parcels of light were absorbed and emitted by blackbodies because of the properties of the bodies themselves, Einstein said that the properties of blackbodies were irrelevant and that light was made up of particles anyway, regardless of whether it was emitted and absorbed by objects. If Planck's model is thought of as being taken from a barrel only in pint portions then in Einstein's theory the beer is already present in the barrel as pint portions, even before it is tapped!

According to Einstein, each photon of light penetrating a metal surface would collide with an electron in the metal and transfer its energy to the electron. If the amount of energy transferred from a particular photon was high enough, it would enable the electron to reach the metal's surface and be ejected, so producing the photoelectric effect (Figure 9).

If the intensity of incident light was increased it would, according to Einstein, mean that there are more photons of a given frequency reaching the metal surface in a given amount of time. Therefore, increasing the intensity of incident light should cause more electrons to be emitted and their energy should be unaltered. This is exactly what happens. Einstein also showed that increasing the frequency of light should increase the energy each photon transfers to the ejected electrons but should have no effect on the number of electrons emitted. This, also, is precisely what occurs in the photoelectric effect.

Exact quantitative measurements of the effect of incident light intensity and frequency on electrons emitted during the photoelectric effect were not available when Einstein published his 1905 paper. However, his mathematical analysis of the phenomenon predicted exactly what results should be obtained. These predictions were the acid test of Einstein's theory. If scientific results hitherto unobtained are correctly anticipated by a theory, scientists can have confidence in the theory. In 1916 the US physicist, Robert A. Millikan (1868–1953), accurately carried out the required experiments on the photoelectric effect. His data

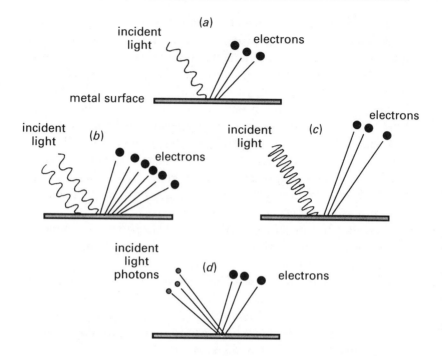

Figure 9.   The photoelectric effect occurs when electrons are
released from a metal surface as a result of light shining on
that surface (*a*). If the intensity of the incident light is increased,
more electrons are emitted (*b*). If the frequency of the incident
light is increased, the same number of electrons is emitted, but
they have a greater energy (*c*). These results are not explicable
in terms of the classical wave theory of light, so Einstein proposed that light can be considered to be made up of particles
(photons), and that each photon collides with an electron and
ejects it from the metal surface (*d*). Increasing the intensity of
the light increases the number of photons (and therefore the
number of electrons emitted), whilst increasing the light's frequency raises the energy of each photon, causing the same
number of higher energy electrons to be emitted.

agreed perfectly with Einstein's predictions. Light did, indeed,
appear to be made up of photons.

   Whilst the wave theory of light did not explain blackbody radiation or the photoelectric effect, the photon idea did not easily

explain interference or diffraction. Indeed, interference and diffraction had been used for many years to support the wave theory. Einstein was aware of the failure of the photon idea to explain diffraction and interference. Nowadays we have come to accept that, at the level of the very, very small, our everyday view of the world does not necessarily hold and that light is both a particle and a wave. This is known as wave–particle duality and applies to all forms of electromagnetic radiation.

When it comes to levels of size as small as photons, both wave and particle aspects can be detected. When light undergoes interference or diffraction it can be considered to be a wave. When it is involved in the photoelectric effect or blackbody radiation, it may be thought of as being particulate. Sir William Henry Bragg (1862–1942), the British physicist, put it more plainly, 'On Mondays, Wednesdays and Fridays light behaves like waves, on Tuesdays, Thursdays and Saturdays like particles, and like nothing on Sundays.'

The wave–particle nature of light was extended to matter by Prince Louis de Broglie (1892–1987), a French physicist, who proposed that not only can electromagnetic waves behave as particles, but also particles can behave as waves. According to de Broglie, even objects as large as a human being or a planet have some wave-like features, although these are minuscule and the particle-like properties dominate. However, electrons are very small particles and de Broglie's theory suggested that electrons should show some wave properties that could be detected. His ideas were subsequently confirmed when electrons were found to show the phenomena of diffraction and interference and their wavelength was measured. Indeed, the wave-like properties of electrons have been particularly useful in the development of **electron microscopes**, which allow small objects to be seen using a beam of electrons instead of a beam of light. Electron microscopes are more powerful than microscopes that use visible light and have been particularly valuable in obtaining images of objects such as viruses and the interiors of living cells, which cannot easily be seen with optical microscopes.

After Einstein had revived Planck's theory of the absorption and emission of light by blackbodies, the Danish scientist, Niels

Bohr (1885–1962), used the ideas of the quantum to propose a new model for the structure of the atom (Figure 10). Bohr explained why atoms of a particular element absorbed and emitted light at specific frequencies. In his model of the atom, negatively

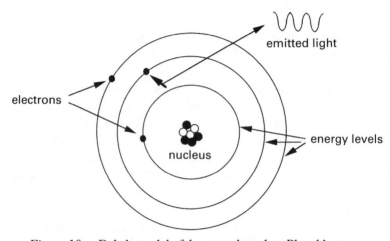

Figure 10.   Bohr's model of the atom, based on Planck's quantum theory. Electrons can occupy only certain energy levels (depicted here as concentric circles) around the atom's nucleus. No intermediate energy levels can be occupied. When an electron jumps from a higher energy level to a lower one, a photon is emitted, the frequency of which is determined by the energy difference between the two levels. In this model, electrons will not simply spiral into the nucleus, and only certain frequencies of light will be emitted when electrons fall from higher to lower energy levels, in agreement with experimental observations.

charged electrons are distributed around a positively charged nucleus. However, the electrons cannot occur just anywhere around the nucleus: they must exist in states of defined energy, called **energy levels**. When an atom absorbs light, the light energy is transferred to an electron and this electron 'jumps' to a higher energy level, but this will occur only if the energy (and therefore frequency) of the light photon is enough to allow the electron to change from the lower to the higher energy level. The electron cannot jump to a place in between the two energy levels: it must be in one or the other level. As a result, the atom absorbs packets (quanta) of energy. Likewise, an atom will emit light energy when an electron jumps from a higher energy level to a lower one, but since only defined jumps can occur the energy of light emitted must have discrete values. Photons emitted and absorbed are therefore of a particular frequency. Bohr's model of the atom showed that the energy levels of electrons are quantised.

Quantum physics, which is required to understand the very small, such as atoms, subatomic particles and electromagnetic radiation, has revolutionised physics. Many physicists are seeking an explanation of the origins of the Universe by combining quantum physics with a theory that allows scientists to understand the very large. One theory that allows the very large to be explained mathematically is Einstein's **Theory of Relativity**, and it is this theory that might perhaps be considered to be the second of the two most important developments in twentieth century physics, quantum physics being the first of these advances.

In view of the tremendous impact that quantum theory had on physics, it is extraordinary that it began with an inspired guess by Max Planck, who simply wanted a mathematical equation that would explain why blackbodies absorb and emit electromagnetic radiation in the way that they do, and that it was strengthened enormously when Albert Einstein sought an explanation for the photoelectric effect.

6

Dr Einstein's fountain pen

E instein's publication on the photoelectric effect in 1905,
which restored the idea that light consists of particles (photons), eventually earned him the Nobel Prize for Physics. However, Einstein is much more well known for his Theories of
Relativity, the first of which was also published in 1905. As a
physicist, there is little doubt that he would have been held with
great esteem by his fellow scientists even if he had not developed
his Theories of Relativity, because his other contributions to
physics were very great indeed. It is a rare event in the history
of the world when a scientist as great as Einstein appears: not
only did he provide answers to many puzzling questions that
classical physics failed to explain, but he also dramatically
changed the way we think about the world around us. Einstein's

relativity also gave birth to modern ideas about the origin of the Universe (Chapter 7). Every physicist alive today is taught in a way that is deeply influenced by Einstein's work and few would doubt that Einstein is the greatest physicist the world has ever seen.

During his life, Einstein became a highly respected and well liked public figure, a situation that is normally somewhat rare for a scientist to achieve. The key to his popularity was probably that he was viewed as a great genius with a kind, unpretentious personality and idealistic political and moral values, and also the fact that his scientific ideas provided a deep insight into the workings of the Universe. He also became involved in several political situations, including the development of the atom bomb; and he was offered the Presidency of Israel, but declined.

Einstein was, like Maxwell and Planck (see Chapters 2 and 5), a theoretical physicist: his experiments were carried out in his mind. (One practical thing he did was to develop and patent, with the Hungarian physicist, Szilard (1898–1964), a noiseless household refrigerator.) When one considers just how much such 'thought experiments' have contributed to science and technology, one is left with the realisation that pure thinking and its mathematical expression are at least as important to human progress as experimental science. Once, when asked by someone if they could see his laboratory, Einstein took a fountain pen from his pocket and said, 'There it is!' On another occasion he commented that his most important piece of scientific equipment was his wastepaper basket, where he threw much of his paperwork containing mathematical computations.

There is a misconception amongst some people that Einstein effectively showed that Newton's Laws of Motion were wrong and some critics of science have suggested that this is evidence that scientists have got it all wrong and that they are constantly contradicting themselves. Newton's First Law of Motion, also known as the Law of Inertia, states that an object will remain at rest or continue in the same direction in a straight line with constant speed unless it is subjected to a net applied force. His Second Law states that an object will accelerate when a net force acts on it, and the net force (F) is equal to the object's mass (m)

multiplied by its acceleration (*a*), that is, $F = ma$. Newton's Third Law of Motion states that when one object applies a force upon another, the second object exerts the same force upon the first, but in the opposite direction. This can be seen clearly in space: when an astronaut throws an object in one direction, the object exerts an equal and opposite force, causing the astronaut to be pushed in the opposite direction.

Some creationists – those who oppose the idea of the evolution of plant and animal species – have even used Einstein's work as evidence that 'facts' (in this case Newton's Laws) established at one time, are sometimes overturned later on. The truth is that Newton's Laws work perfectly well under everyday conditions and they have been tremendously useful to physics and technology. What Einstein did was to show that Newton's Laws of Motion need to be modified when objects approach the speed of light (300 000 kilometres per second; 186 000 miles per second). At speeds normally encountered in daily life, which are a tiny fraction of the speed of light, moving objects behave in such a way that Einstein and Newton's descriptions of them become indistinguishable. Indeed, Einstein really refined Newton's Laws of Motion so that they worked at all speeds: this modification of existing ideas is the normal way by which science progresses.

Einstein revolutionised the whole of physics, particularly with his work on Relativity, which involves two main theories: the Special Theory of Relativity, which he published in 1905, and the General Theory of Relativity, which he published in 1916. Before these are described, some knowledge of the scientific context of Einstein's ideas is required.

Physics before Einstein: the ether

Prior to the twentieth century, when the notion that light was made up of waves dominated physical thought, scientists believed that there must be a medium in which light waves were propagated. After all, water waves were carried by water and sound waves were carried by molecules in the air (or whatever the medium was that transmitted the sound). Sound, for example, does not travel through a vacuum, because there are no molecules to transmit it. Waves, it was thought, could not be transmitted without there being some substance to carry them, so light had to have a supporting medium. The medium that allowed propagation of light waves was called the **ether**. (The term 'ether' is derived from the Greek word for 'blazing': the ancient Greeks used the word to describe the element from which the stars were made.) Nobody knew exactly what the ether was: it was necessary to assume its existence in order to accept the view that light was made of waves.

Light also has a finite speed: it takes some time for it to travel from its source to another object. The light from distant stars that arrives on Earth has travelled so far that what we detect is light that actually left a star thousands, millions or even billions of years ago, depending how far away the star is. Of course, that light has a finite speed is not noticeable in everyday events: it is so fast that it takes only a billionth of a second for light to travel to our eyes from a source a few metres (or feet) away. When a light is turned on in a room we do not notice the tiny fraction of a second that it takes for the light to travel from the bulb to our eyes. The first attempt at measuring the speed of light was made in 1676 by the Danish astronomer, Olaf Roemer (1644–1710). More accurate measurements were made later on, and it became clear that light did have a definite speed. According to Maxwell's equations, which assumed that light consisted of waves (Chapter 3), light should travel more slowly in denser transparent materials such as glass and water than it does in air. Experiments in which the speed of light was measured in different materials confirmed this. For example, the speed of light is reduced by about a third, to 200 000 kilometres per second (124 000 miles per second) when it enters a block of glass, largely because of the time it takes to be absorbed and emitted by atoms in the glass during its transmission.

When more of the detailed properties of light were established, calculations showed that the ether had to be a rigid solid that vibrates when light passes through it. Since nobody could see or detect the ether, it had to be very fine as well as rigid, and it had to be everywhere, even in a vacuum, because light can travel through a vacuum. Clearly, the ether, if it existed, had to be a new kind of 'substance' compared with those previously known.

Nevertheless, if the ether existed it should be possible to devise experiments that could detect its presence. Indeed, in 1887 two US scientists, Albert Michelson (1852–1931) and Edward Morley (1838–1923), carried out a precise experiment aimed at detecting the ether. Their reasoning was that since the Earth is in motion it should be moving through the stationary ether in a similar way to a swimmer moving through water. This should create an 'ether pressure' that would be due to the ether apparently 'flowing' past the Earth. Furthermore, because the Earth is rotating about its own axis as well as travelling around the Sun in an elliptical orbit, its direction of travel through the ether will vary at different times, and so the direction of the 'ether pressure' at a particular point on the Earth will change with the Earth's motion. If the speed of light was measured from a point (a physics laboratory, for example) on the Earth, its measured speed would depend on whether the light was transmitted in the same or a different direction compared with that of the Earth's motion in the ether. Calculations showed that the measured speed of light travelling, on a round trip, parallel to the direction of apparent flow of the ether should be slower than the measured speed of light travelling at right angles to this direction of flow. In much the same way, a person who swims up and down a river against the water's current will take longer to cover a certain distance than he or she would take to travel the same distance by swimming at right angles to the current.

Michelson and Morley used a very accurate device for simultaneously measuring the speed of light in different directions compared with the Earth's motion and their equipment should easily have detected any movement through the ether. However, they failed to find any differences between their measured values of the speed of light, even after thousands of attempts. They could not detect any ether pressure – they could not detect the ether.

Since Michelson and Morley carried out their experiments, many more scientists have attempted to find evidence for the ether. Some of these experiments were carried out more recently using highly sophisticated technology. None of them has found any evidence for the existence of an ether: the speed of light was always the same whenever and in whatever direction it was measured.

To the physicists of the nineteenth century, the ether was not only an essential medium in which light waves were propagated: it was also an absolute frame of reference that allowed the true motions of objects to be determined. To illustrate the significance of this statement, consider a train moving through the countryside at a certain constant speed. Passengers on the train have no doubt that they are moving because they can see the trees and fields go by. However, if the background trees and fields are removed, it becomes more difficult to assess whether or not one is moving, unless the train changes its speed, when a passenger can feel the force produced by the slowing down or speeding up of the train. If another train appears on an adjacent track, it becomes difficult to assess which train is moving the fastest, and even whether or not one train is stationary. If one train was travelling at 100 kilometres per hour (65 miles per hour) and the other was moving in the same direction at 60 kilometres per hour (40 miles per hour), the situation would look much the same to passengers as that existing when one train travelled at 40 kilometres per hour (25 miles per hour) with the other train at rest. In addition, without the background as a frame of reference, it would not be possible for passengers of either train to be sure which train was moving: each of them could claim that they were at rest and the other was moving, and each could claim that they were moving and the other was at rest.

The point about the background scenery of a moving train is that it provides a frame of reference from which one can say that one is in motion. It allows one to say that the train is moving *relative* to the scenery (the Earth). Without some frame of reference, it is impossible to say whether anything is moving or at rest, provided that its speed is constant. Indeed, the Earth is revolving around the Sun at 108 000 kilometres per hour (67 000 miles per hour); the Sun and its solar system are revolving around the centre of our Milky Way galaxy at 500 000 kilometres per hour

(310 000 miles per hour); and our galaxy is travelling through the Universe at a speed of 2 300 000 kilometres per hour (1 500 000 miles per hour). There is no doubt that we are all whizzing through the Cosmos at a breathtaking speed, but we fail to perceive this in our everyday lives.

The ether provided physicists with a universal background 'scenery' against which the absolute motions of objects could be assessed. When we say that the Earth is revolving around the Sun, we use the Sun as a frame of reference: the revolution of the Earth is considered relative to the Sun. However, if the Sun is moving as well, we get no indication of the absolute motion of the Earth and we have to consider the motion of the Sun relative to something else, for example, another star. But if this other star is also moving, we need another frame of reference with which to measure its speed, and then what is the true (absolute) speed of the Earth or the Sun? The stationary ether provided that frame of absolute rest: it was considered to be completely stationary and not to be moving with the Earth, the Sun, the stars, or any other celestial object. If the speed of the Earth could be measured with respect to the ether, then its absolute speed would be determined, and the speed of anything measured relative to the Earth could then be used to determine the absolute speed of that object.

Newton was well aware of the problem of relative motion. His concept of absolute rest was a religious one. Although we cannot be sure that any of the objects we see are absolutely still, he said that there *is* such a thing as absolute rest and it is known by God. Although faith can be a good thing under some circumstances, it does not provide any practical way of sorting things out in the physical world. The ether was a more concrete and tangible way of providing a frame of reference that was at absolute rest. Unfortunately, its existence could not be detected, and absolute rest was on shaky ground.

This was a considerable blow to classical physics. For example, Newton's Laws of Motion required the existence of a frame of reference that was at absolute rest, and Maxwell's equations also needed an absolute frame of reference. Without the ether, there could be no medium for transmission of light waves and there could be no absolute frame of reference against which to measure the speeds of moving objects.

In order to explain Michelson and Morley's failure to detect the ether, the Irish physicist, George FitzGerald (1851–1901), proposed that a moving object shortens in length in the direction of its absolute motion. According to FitzGerald, a ruler pointing in the same direction of the Earth's motion would contract lengthwise, whereas a ruler pointing at right angles to the Earth's motion would not contract lengthwise (although it would contract widthwise, since its edge is moving in the same direction as the Earth). FitzGerald derived a mathematical equation showing that any measurements made of the speed of light in the same direction as the Earth's motion would be compensated for by contraction of the measuring apparatus and would be the same as the measurements of the speed of light made at right angles to the Earth's movement. In other words, the ether could still exist but it might not be detected by measuring the speed of light as Michelson and Morley did. The idea that objects get shorter in the direction of their motion was a rather peculiar one, but it did provide a possible answer to the experimental data; and Einstein used the same idea in his Theory of Relativity.

The degree of contraction of an object moving under most everyday circumstances was minuscule, according to FitzGerald, but as the speed approached that of light, contraction became obvious. Thus, a 30 centimetre (1 foot) ruler would contract to about 27 centimetres (10.5 inches) when it was travelling at half of the speed of light (150 000 kilometres per second); at three-quarters of the speed of light it would become about 20 centimetres (8 inches) long; and at the speed of light it would not have any length at all!

Subsequently, the Dutch scientist, Hendrik Antoon Lorentz (1853–1928), showed mathematically that not only should an object contract in its direction of absolute motion, but also its mass should increase. For example, an object weighing a kilogram at rest should weigh about 1.15 kilograms when it travelled at half the speed of light; its mass would increase to 1.5 kilograms at three-quarters of the speed of light; and at the speed of light its mass should become infinite! Lorentz described this effect for charged particles in motion, but Einstein later showed that the mass increase occurred with all moving bodies.

It may sound ridiculous that a moving object not only contracts but also that it gains mass as its speed increases: the whole idea seems to be contrary to common sense. However, common sense deals with the everyday world, where speeds are very small compared with that of light. A car travelling at 60 kilometres per hour (40 miles per hour) is moving with a speed of only one eighteen-millionth of the speed of light and this results in a contraction and mass increase that are not noticeable. As speeds approach that of light, contraction and mass increase do become apparent, but we never see objects moving at these speeds under most normal circumstances.

Although FitzGerald had proposed contraction to save the existence of the ether, neither contraction in the direction of an object's motion nor its mass increase with speed were dependent on the existence of the ether. Albert Einstein arrived at his **Special Theory of Relativity** by ignoring the ether: according to him, the ether was not needed. This interpretation of the motion of objects also solved many of the problems that classical physics had failed to explain.

The Special and General Theories of Relativity

Einstein's Special Theory, and later his General Theory, were such monumental contributions to physics that many biographers have tried, with little success, to explain what made him such a genius. Einstein was born in 1879 in Germany, but he later renounced his German citizenship and became a Swiss citizen. Very little in terms of his childhood could have led anyone to think that he would have made such a gigantic intellectual contribution. Indeed, he did not even speak properly until he was three years old, and his parents were concerned that he might be having learning difficulties! He was a solitary child who later returned to isolation: 'The individual who has experienced solitude will not easily become a victim of mass suggestion', he once said.

Einstein hated regimentation of any kind, including the rote learning of the type that occurs at school. This made him rebellious of schoolteaching methods, but it also encouraged him to be self-taught and to think independently. Mathematics was his best subject and he later studied the subject at the Swiss Federal Polytechnic School in Zurich. In 1902 he obtained a position as patent clerk at the Swiss Patents Office in Berne. While he worked there he did theoretical physics in his spare time and published his earth-shattering ideas on the photoelectric effect (Chapter 5), Brownian motion, and the Special Theory of Relativity.

The Special Theory of Relativity was Einstein's theoretical way of solving some of the problems associated with classical physics that seemed to be contradictory. For example, according to classical physics, charged particles are associated with a magnetic field when they are moving. The question arises: moving with respect to what? A charged particle on the surface of the Earth that is stationary with respect to the Earth is still moving, since the Earth is moving. The mathematical equations that described magnetic fields associated with charged particles required an absolute frame of reference in order to say whether or not a particle was moving. Einstein eliminated the need for this absolute framework and his Special Theory, which was a theoretical study of objects moving at a constant speed, removed many other apparent contradictions of classical physics. In Einstein's analysis of motion and electromagnetism, there is no need for an absolute frame of reference: the laws of physics are the same for any observer, regardless of his or her motion.

Newton's Laws of Motion allowed an object to travel at any speed. Provided that a sufficiently large force could be exerted on the object for a sufficient length of time, there was nothing to stop it from reaching speeds as high as, or even greater than, the speed of light. However, one of Einstein's most important premises was that the measured speed of a uniformly moving object can never be greater than the speed of light. His equations, and those of FitzGerald and Lorentz, predicted zero length and infinite mass of an object travelling at the speed of light, and at speeds greater than that of light the mass and length of an object cease to have any meaning – they become imaginary. Einstein already had some support for this idea: nobody had ever observed any object moving faster than light. Indeed, this statement is still true today.

Einstein also said that the speed of light in a vacuum will always be measured to be the same value, regardless of the speed of the light source or the observer measuring it. To understand the significance of this proposition, consider a stone that is thrown from a moving train. Its speed (measured by an observer on the ground in the background scenery) will be greater if it is thrown in the direction the train is moving than if it is thrown in any other direction. The speed with which the stone is thrown is added to the speed of the train if it is thrown in the direction of the train's motion, whereas the stone's speed is subtracted from the train's speed if it is thrown in the exact opposite direction (Figure 11). Light, on the other hand, behaves rather differently. Consider a torch being shone from a moving train. Instead of travelling faster if it is shone in the direction of the train's movement than if it is shone in a different direction, the light travels at the same speed in any direction. Even if the train moves at half the speed of light, the measured speed of a light beam that is shone in the direction of the train's motion is the same as it would be if the train were not moving, instead of being one and a half times that speed (Figure 11).

A consequence of these assumptions about the speed of light is that it is impossible to calculate the absolute speed of an object. Einstein saw no use for an ether against which to measure absolute speed. Everything could be explained without an ether, and this certainly would agree with the fact that nobody had ever detected it.

If the measured mass of an object increases as it approaches the speed of light, what, then, becomes of the Law of Conservation of Mass (that matter cannot be created or destroyed, see Chapter 10), which, at the time the Special Theory was formulated, was widely accepted as fact? Surely an increase in mass means creation of matter, in contradiction of this law? Einstein arrived at an answer to this problem when he derived his famous equation, $E = mc^2$. This equation shows that the energy (E) of an object is

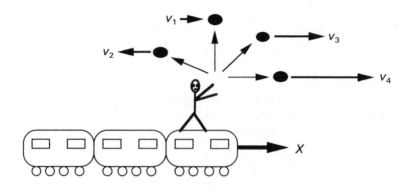

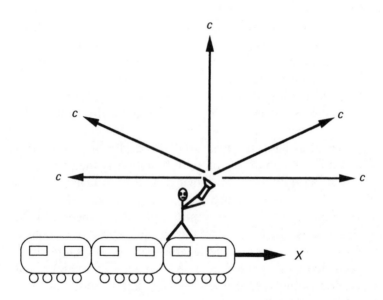

Figure 11. Constancy of the measured speed of light. Whereas the speed of a stone (top, v_1–v_4) thrown from a moving train depends on the direction in which it is thrown as well as the speed (X) of the train, the speed of light (bottom, c) is found to be the same irrespective of the direction in which it is projected and regardless of the speed of the train. The arrows represent the direction, and the lengths of the arrows represent the speed.

related to its mass (*m*) and the speed of light (*c*). Einstein reformulated the Law of Conservation of Mass and transformed it into the Law of Conservation of Mass–Energy: matter and energy cannot be created or destroyed, but they can be converted one into the other.

Einstein's Special Theory immediately provided an explanation for radioactivity, which had been a puzzle to physicists (Chapter 4). Now it could be seen how seemingly endless amounts of energy were given off by radioactive atoms: small amounts of matter in the nucleus of the atoms were being converted to energy. The Law of Conservation of Mass–Energy was being obeyed. Indeed, after Einstein's Special Theory of Relativity was announced, physicists searched for and found the decrease in mass of radioactive nuclei that was required to produce radio-active energy. The equation, $E = mc^2$, indicates that when an object is converted to energy, the amount of energy produced is equal to the mass of the object multiplied by the speed of light, multiplied again by the speed of light. Because the speed of light is very large, this means that a small quantity of matter can be converted into an enormous amount of energy. This is the principle of nuclear power and atom bombs: small amounts of matter from nuclear reactions are converted into large amounts of energy. In view of the fact that Einstein was an overt idealist and pacifist, it is ironic that the destructive use of the matter-to-energy conversion provided some of the strongest evidence for his theory.

Einstein's Special Theory of Relativity also provided some novel interpretations of the concept of time. Newton had ascribed a knowledge of absolute time to God, just as he had done with an absolute frame of reference for motion. A consequence of the constant, finite speed of light was that measurements of time, in addition to those of length and mass, also vary according to the motion of an object. Einstein's mathematical equations showed that as a body approaches the speed of light, time appears to lengthen. For example, to an object moving at ninety-eight per cent of the speed of light, one second becomes five seconds long. This phenomenon is called **time dilation**. As with changes in measured length and mass, the effects of speed on time are insignificant at speeds encountered in ordinary daily life, but they do become significant as the speed of light is approached.

Evidence for time dilation has been forthcoming from several sources. For example, nuclear accelerators exist that allow sub-atomic particles to attain very high speeds, approaching that of light. Some of these particles are unstable and break down at a defined rate. As they move faster, the particles break down more slowly: the time taken for them to decay becomes longer as they approach the speed of light. Another piece of evidence that supports Einstein's contention that time slows down as objects move faster comes from measurements of the rates of ticking of highly sophisticated atomic clocks. Such clocks have been shown to run more slowly on fast moving jet planes and on satellites than similar clocks on Earth.

The Special Theory of Relativity deals with the effects of uniform motion on length and mass of an object and on time. Einstein later developed his General Theory of Relativity, which is concerned with non-uniform motion and proposes a new way of looking at gravity. To Newton, gravity was a force of attraction between two objects: large objects exerted a greater gravitational pull than small ones. Einstein did away with this idea. He showed that gravitational attraction was equivalent to acceleration (increase of speed); this was called the **Equivalence Principle**. An object in an upwardly accelerating container is pulled down to the floor of the container in exactly the same way as an object is pulled towards the Earth by gravity.

To Einstein, gravity was not a force between objects, but a curvature of space–time that was caused by the presence of mass. An object does not fall to the Earth because it is being attracted by the Earth's gravitational pull: it falls because it takes the easiest route along the curvature of space–time created by the Earth's mass. An analogy is that of a ball-bearing placed on a sheet of elastic material. The ball-bearing (Earth) creates a depression in the sheet (space–time), and the sheet is curved around the ball-bearing, becoming more steeply curved nearer the ball-bearing (Figure 12).

The General Theory of Relativity was testable, like all good theories. It predicted, for example, that light should be bent by

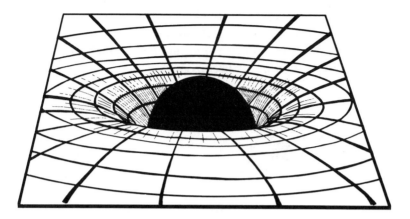

Figure 12. Gravity as space–time curvature. Einstein considered gravity to be caused by matter influencing the curvature of space–time, rather than as a force of attraction. An analogy is that of a ball-bearing in the middle of an elastic sheet, causing the sheet to curve in its vicinity. A small spherical object rolled across the sheet would fall towards the ball-bearing because of the curvature of the sheet; in a similar way, objects 'fall' to Earth under the influence of the Earth's gravity.

gravity. Light passing close to the Sun would, according to Einstein, be bent by the curvature of space–time caused by the Sun. This should be measurable under the correct circumstances. The best situation in which to observe this bending of light by the Sun is a total eclipse of the Sun. In 1919 such an eclipse occurred off the coast of West Africa, and a group of British scientists organised an expedition to test Einstein's theory by measuring the positions of stars in the neighbourhood of the Sun and looking for the predicted bending. The results were in agreement with Einstein's theory, and this caused a sensation that made Einstein a household name around the world. When Einstein received a telegram informing him of this confirmation of General Relativity, he was remarkably calm about it. One of his students who was with him at the time commented on how exciting the event was, but Einstein apparently responded with little emotion and said, 'But I knew that the theory was correct'. When asked what he would have thought if the results had disagreed with his theory,

Einstein replied, 'Then I would have felt sorry for dear God, because the theory is correct.' However, a letter written by him at the time to his mother does indicate that he really was excited by the confirmation of his prediction.

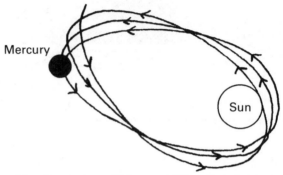

Figure 13.   Precession of Mercury. When it undergoes a revolution around the Sun, Mercury precesses: it does not complete a closed ellipse. This phenomenon was not entirely explicable in terms of classical physics, but Einstein's General Theory of Relativity gave a precise explanation for it.

Although many physicists later doubted the accuracy of the measurements made during the 1919 solar eclipse, numerous other observations have since supported Einstein's idea about gravity. In addition, Einstein knew that his General Theory provided an answer to a puzzle, that of the peculiar orbit of the planet, Mercury, which classical physics could not explain. It had been known since the mid nineteenth century that the orbits of the planets around the Sun are nearly, but not quite, closed ellipses. The planets precess; that is, they return to a slightly different place after each orbit, producing an orbit consisting of loops rather than a closed ellipse (Figure 13). Mercury – the planet nearest to the Sun – precesses to the greatest extent. When Newton's theories were applied to Mercury's precession, they could not explain it properly. The phenomenon baffled many physicists, especially because no other planet close to the Sun was found that might be exerting a force on Mercury and causing the anomaly. Classical physics simply did not explain this orbit. However, General Relativity explained it exactly: when Einstein applied his equations describing the effects of the Sun on the curvature of space–time, Mercury was predicted to have the precise orbit that was found.

Einstein's General Theory not only provided a brand new way of looking at Nature and explaining properties of the Universe, but also gave birth to modern cosmology, the study of the origins and properties of the Universe. His equations of General Relativity predicted that the Universe was not static – it was either expanding or contracting. Although Einstein rejected this idea, believing that the Universe was static, other physicists used his equations to propose models of an expanding Universe, and this gave rise to the Big Bang theory for the origin of the Cosmos. Einstein's 'thought experiments' had clearly revolutionised our understanding of existence, and will continue to do so.

7

The Big Bang, or how it all began

E verything in the Universe – the hundreds of billions of galaxies, all of the planets, every pebble, grain of sand, man,
woman, child, animal, plant, bacterium, indeed every atom and
bit of energy – were once compressed together in an object much,
much smaller than the full stop at the end of this sentence.
Ridiculous though this sounds, it is part of a theory that cosmologists have for the origins of the Universe, and which many take
very seriously. This theory, and variants of it, constitute the so-
called **Big Bang** theory of the origin of the Universe.

Religions say nothing concrete about how the Universe began.
Much that is said either conflicts with the available scientific
evidence or is accepted simply on the basis of faith. For example,
in 1650 Archbishop Ussher of Ireland (1581–1656) calculated,

on the basis of his readings of the Bible, that the Earth was created at 9 am on the morning of 26 October 4004 BC, but we are certain now that the Earth is well over four billion years old (Chapter 9), and the Cosmos is unquestionably older than the Earth. Scientists make what observations they can within their sphere of existence and they make logical deductions based on the evidence. The Big Bang theory for the origin of the Universe provides the clearest explanation of the scientific data that makes sense.

The origin of the Universe is a fundamental question. There cannot be many people who have not questioned where everything came from. The Cosmos is vast – too large for the imagination to comprehend – yet as one goes back in time to its beginning, everything becomes simplified and the Universe becomes easier for the mind to grasp. It has been said that humankind will have a more complete understanding of the origins of the Universe than of a living cell. It is as if the products of the evolution and differentiation of the Universe are far more complex than the Universe itself just as an egg is simpler than the animal into which it develops. Whether any moral values can be gleaned from the Big Bang theory remains to be seen. What is clear is that every atom in the Cosmos had the same origin, and once upon a time everything was as one.

Until the twentieth century, scientists generally considered the Universe to be infinite and static. However, several discoveries together strengthened the idea that the Universe is not static and that it has expanded over a period of fifteen to twenty billion years from an infinitesimally small and hugely dense point. Modern cosmology began with Einstein, whose General Theory of Relativity provided a mathematical model that predicted a non-static Universe. Unfortunately, even Einstein was prone to adhering to some established beliefs, and he changed his mathematical equations, forcing them to fit a static Universe. It was another ten years before evidence was obtained that the Universe is indeed expanding and the idea of a non-static Universe was taken seriously by physicists.

Once it became clear that the mathematical equations of theoretical physicists could provide models of the Universe and that many of these models could be tested by observing the properties of the Universe from Earth, physicists realised that the secrets of the Universe, its structure, behaviour and origins were quite within their grasp, and cosmology became a science in its own right.

Early ideas about the Universe

Before Einstein came up with the General Theory of Relativity in 1915, the majority of physicists believed that the Universe was infinite and had existed essentially in its present form since it was created. One puzzle, upon which several physicists pondered, was later called **Olbers' paradox**, after the German astronomer, Heinrich Olbers (1758–1840). Olbers' paradox addresses a problem associated with the idea that the Universe is uniform and infinite. If the Universe does go on forever in space and its stars are evenly distributed, why is the sky dark at night? Surely, every line of sight from the Earth would contain a star if it were extended far enough in space, and therefore light should reach the Earth along all lines of sight, which would make the sky completely bright? Some reasonable explanations were proposed to explain the darkness of the night sky. For example, several physicists suggested that the Universe was very young and that light had not yet reached us from the most distant stars, so that the dark regions of the sky correspond with these distant realms of the Universe. Others believed that the space in between distant stars and the Earth blocked out the light and prevented it from reaching us. Few physicists questioned the idea that the Universe was infinite: Newton's laws of gravity clearly indicated that a finite and static Universe would collapse upon itself because of the gravitational pull of the inner objects on the more peripheral ones.

When Einstein applied General Relativity to the structure of the Universe, he did not question the idea that the Universe was static. As his mathematical equations initially suggested that the Universe was either expanding or contracting, Einstein decided to alter them by adding a 'cosmological constant', which allowed the mathematics to conform to an eternally homogeneous and static Universe. Other scientists left out the cosmological constant and proposed models of an expanding Universe, and in the 1920s, Edwin Hubble (1889–1953) gave credibility to these theoretical deductions by obtaining experimental evidence for an expanding Universe. Einstein can be regarded as the Father of Modern Cosmology and perhaps the world's greatest ever scientist, but he failed to be the first to propose an expanding Universe. He later said that the addition of the cosmological constant to his mathematical equations was 'the biggest blunder of my life'.

An expanding Universe

The Dutch physicist, Wilhelm de Sitter (1872–1934), was the first person to propose a model for an expanding Universe based on Einstein's equations of General Relativity, but his equations were valid only if the Universe contained no matter. Other physicists used de Sitter's model to develop mathematical models that would describe what would happen if his Universe did contain particles of matter. They came to the conclusion that these particles of matter would all be moving away from each other. During the 1920s, the Russian physicist, Alexander Friedmann (b. 1888) and the Belgian scientist, Georges Lemaitre (1894–1966), examined Einstein's equations of General Relativity, making the assumption that the Universe was homogeneous and that it looked the same from whatever direction it was observed. Lemaitre's interpretation was that if the galaxies were moving apart, they should have been closer together in the past, and if one extrapolates far enough back in time, there must have been a stage at which all of the matter in the Universe had been concentrated into a single point with a very high density. This was the birth of the Big Bang theory. In fact, the term 'Big Bang' was originally used by one of the theory's opponents as a derogatory term, but it eventually gained acceptance by its proponents. Lemaitre also predicted the existence of the **red shift**, which was soon supported experimentally.

During the early part of the twentieth century, astronomers, and in particular the US scientist, Vesto Slipher (1875–1969), measured the motions of celestial objects called nebulae. The majority of known nebulae showed a red shift: the light from them that reached the Earth consisted of electromagnetic waves shifted towards the red end of the spectrum compared with what would be expected for a static light source. This red shift meant that these nebulae were moving away from the Earth. So, it appeared, almost all nebulae are moving away from us.

Hubble and his colleague, Milton Humason (1891–1972), made an important advance in their measurements of red shifts of nebulae: they showed that the further away the nebulae were from Earth, then the greater was their red shift, and hence their velocity. Hubble's measurements of the distances of nebulae from the Earth established the notion that nebulae were well outside of our own Milky Way – they were, in fact, other galaxies. Hubble's Law, which relates the red shifts to the distances of galaxies from Earth, provided evidence for the idea that the Universe is expanding. An analogy is a currant bun: when the bun expands whilst it is baking, the currants get further apart from each other, regardless of their positions in the bun (Figure 14). Likewise, if the Universe is expanding, then all galaxies should be moving away from each other: Hubble and Humason's data agreed with the models proposed for an expanding Universe.

Although Hubble and Humason's data did indicate that the Universe is expanding, there was a problem associated with the age of the Universe that their data predicted. Their calculations suggested that the Universe was a mere two billion years old, but this was in conflict with strong evidence, from accurate studies of the oldest rocks, that the Earth was about four and a half billion years old, and later on the conflict deepened when some stars and galaxies were deduced to be ten billion years old. This problem was eventually solved: it was due largely to inaccuracies that Hubble and Humason had encountered in measuring distances between the Earth and galaxies: they had underestimated them. The interpretations they made about the expanding

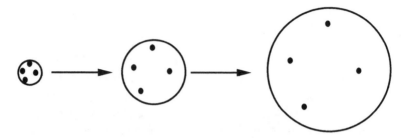

Figure 14. The expansion of the Universe can be compared loosely with an expanding currant bun. As the bun expands during cooking, the currants all move apart from each other. From the point of view of one currant, all the other currants are moving away from it. This is similar to the movement apart of the galaxies in the Universe, and explains why the vast majority of galaxies show a red shift.

Universe were still correct, but now the age of the Universe that was predicted by their data tallied with data obtained from other sources.

Two major models were then propounded based on an expanding Universe. The first one stated that the Universe was expanding indefinitely. The second model said that the Universe would reach a critical state of expansion at which it would start to contract again, eventually producing a 'Big Crunch', when all matter would be compressed into a minute point of high density. The second model suggested that the Universe undergoes cyclical periods of contraction and expansion.

Once the Big Bang theory had gained some credibility, physicists began to examine the theoretical structure of the Universe at various points in time, going back to fractions of a second after the Big Bang. For example, the temperature of the Universe at different times after the Big Bang could be calculated mathematically, and it was already known from experiments what states of matter and energy would exist under some of these conditions. From these attempts at working out the composition of the early Universe, predictions were made that allowed the Big Bang theory to be tested experimentally. One of these predictions, which was made by the Soviet–US physicist, George Gamow (1904–1968), was that the early Universe was dominated by very hot radiation. When the Universe was one second old, it had a temperature of ten billion degrees Centigrade, whereas three minutes later, it was only one billion degrees Centigrade. By the time the Universe was a million years old, its temperature had dropped to about 3000 degrees Centigrade. Gamow's colleagues, Ralph Alpher (b. 1921) and Robert Herman (b. 1914), calculated that the radiation of the early Universe should have permeated the whole of the Cosmos and by now it would have cooled down to -268 degrees Centigrade. This radiation left over from the early Universe should still exist, therefore, and might be detectable: it was predicted to have a frequency in the radio/microwave frequency range.

Gamow and Alpher proposed a model of the expanding Universe that explained how the elements, hydrogen and helium, had formed some time after the Big Bang from smaller, subatomic particles. Of the ninety-two naturally occurring elements, hydrogen and helium are by far the most abundant in the Universe. Hydrogen accounts for nearly three-quarters of the matter in the stars and galaxies, and nearly twenty-five per cent comes from helium. Hydrogen and helium are the two lightest elements, and they were the first elements to form in the early Universe when it expanded and cooled down. Prior to that, the temperature of the Universe was too high to allow atoms to exist. Heavier elements occur in very small amounts in the Cosmos compared with hydrogen and helium. Gamow and his colleagues could not explain, on the basis of the Big Bang theory, the heavier elements, but they could explain very accurately the relative levels of hydrogen and helium. Most heavier elements are now thought to be made in the stars.

Gamow was known for his great sense of humour, and he added an extra name – that of the physicist, Hans Bethe (b. 1906) (pronounced 'beta') – to the scientific paper that he and Alpher published on their model of the early Universe. This meant that the three authors had names that sounded like the first three letters of the Greek alphabet (alpha, beta and gamma).

The prediction of the abundances of hydrogen and helium provided support for the Big Bang theory, adding to the evidence for expansion that was supplied by Hubble and Humason. However, the radiation left over from the early Universe that was predicted by Alpher and Herman would, if it could be found, provide even stronger evidence for the Big Bang theory.

Although the cosmic radiation left over from the primordial Cosmos was predicted in the 1930s, few cosmologists took it seriously, and there was little effort made to detect the radiation. The background microwave radiation that was, indeed, the echo of the early Universe, was not discovered until the 1960s. Even then, the scientists who discovered it – Arno Penzias (b. 1933) and Robert Wilson (b. 1936) – did so serendipitously. It earned them the Nobel Prize for Physics in 1978 and provided the best evidence yet for the Big Bang theory.

The cosmic microwave background radiation

The US physicists, Robert Dicke (b. 1916) and James Peebles (b. 1935), who worked at the Holmdel Laboratories of Princeton University in New Jersey, were interested in searching for the predicted radiation that was left over from the Big Bang. With their colleagues, they were in the process of building a detector that might allow them to discover this radiation, when Penzias and Wilson came up with the discovery without even having any intention of finding the cosmic background radiation.

Penzias and Wilson were working at the Bell Laboratories in New Jersey, only about twenty-five miles away from Dicke and Peebles' laboratory. The Bell Laboratories were keen to develop their satellite communication systems, and one of their aims was to transmit information at microwave radiation frequencies. For this purpose, they had developed a large horn-shaped receiver, nearly 7 metres (23 feet) long, which detected signals bounced off balloons high in the Earth's atmosphere.

Penzias and Wilson were more interested in using the large horn-shaped receiver for studying radio astronomy than for developing microwave communication systems. They were allowed to use the receiver for their studies of radio waves coming from the Cosmos. However, there was considerable interference with their apparatus by background microwaves, so they set about getting rid of this unwanted radiation. They managed to eliminate some of the interference, but noticed that there was some left that steadfastly refused to go away. Even after clearing the horn antenna of pigeon droppings, which they thought could be causing this interference, Penzias and Wilson still could not rid themselves of the annoying microwave background radiation. Wherever they pointed their antenna, the interfering radiation was there, and it was equally strong in all directions. Getting rid of it proved to be an impossible task, and they were beginning to accept failure.

Then, in 1965, Penzias and Wilson decided to seek advice about their problem. Someone suggested they contact Robert Dicke at the Holmdel Laboratories for help in solving their problem. Penzias telephoned Dicke to ask for his advice on the microwave noise, and the four scientists, Dicke, Peebles, Penzias and Wilson, soon got together to discuss the matter. It became clear that the noise was, in fact, the prize that Dicke and Peebles were seeking: it was the cosmic radiation left over from the Big Bang. What Penzias and Wilson had thought might be interference from pigeon faeces turned out to be far more significant – it was the whisper of the early Universe! The temperature of the background radiation detected by Penzias and Wilson was −269.5 degrees Centigrade, which was not far off that predicted by Peebles and Dicke, and earlier by Alpher and Herman.

Before they spoke with Peebles and Dicke, Penzias and Wilson had no idea that their microwave background interference would provide the strongest evidence yet for the Big Bang theory of the Universe. Careful measurement of the background cosmic microwave radiation by Penzias and Wilson showed that it behaved as expected for blackbody radiation (see Chapter 5), which also had been predicted to be a feature of the radiation left over from the Big Bang. Subsequent studies by other scientists confirmed that this radiation is, indeed, the remains of the cosmic event that gave birth to the Universe. The photons that make up the cosmic microwave background are the oldest ones in the Cosmos; they have been around for more than ten billion years.

If the Big Bang theory is correct, and there is every reason to believe that it is, since evidence for it has been forthcoming from various sources, how, then, can we form a picture of the origins of the Universe and its evolution into today's Cosmos?

The aftermath of the Big Bang

The mathematical equations that describe the Big Bang theory are meaningful only after one ten-million-trillion-trillion-trillionth of a second (10^{-43} second) after the 'moment' of creation. Before this time, which is called the **Planck time**, the laws of physics as we know them are meaningless: General Relativity does not apply to this stage of the Universe's existence. At the 'moment' of creation, before this time, the Universe was, according to General Relativity, infinitely dense and space–time was infinitely curved: such a point is called a **singularity**.

However, predictions can be made of the state of the Universe after the Planck time. From the time of the Big Bang until the Planck time, all of the four forces of Nature – gravitational, electromagnetic, weak nuclear and strong nuclear forces – were unified into one. After this time, the forces began to appear as separate entities.

According to one Big Bang model, the Universe expanded from a tiny fraction of the size of an atom to an object the size of a tennis ball within the first trillion-trillion-trillionth of a second. When it was a trillionth of a second old, it had expanded to a few metres in diameter. A measure of its rate of expansion can be seen from its size when it was a mere billionth of a second old: at this time it was almost a quarter of a million times bigger than the Earth – as big as our solar system. By the time the Universe was one second old, it had cooled down to ten billion degrees Centigrade. It was still very hot – several thousand degrees Centigrade – even when it was a few thousand years old.

Atoms could not be formed until the Universe was a million years old. Before that, subatomic particles were constantly colliding with radiation photons. After a half a million to a million years, photons were free to move throughout the Universe, and the cosmic microwave background radiation discovered by Penzias and Wilson is actually the cooled remains of this radiation.

The Big Bang was not really an explosion. For one thing, there would have been no noise associated with it, because sound waves could not have existed. In addition, the Big Bang did not occur *in* space or time: when we say that the Universe expanded, we mean that space–time itself expanded. We cannot meaningfully discuss what was beyond the 'edge' of the expanding early Universe.

If the early Universe was homogeneous, how did the galaxies form? After all, today's Universe is 'particulate' with galaxies dotted throughout, and there is evidence that galaxies cluster in enormous numbers called 'great walls' in some parts of the Universe. In order to give rise to this heterogeneity, the early Universe could not have been completely uniform, but fluctuations should have occurred. Because atoms began to form when the Universe was a half a million years old, and this was the time when radiation became free to move throughout the Universe without colliding with subatomic particles, many physicists believed that there must have been fluctuations in the Universe at this time, and that these inhomogeneities gave rise to local regions in which today's galaxies were formed. Physicists predicted that these fluctuations in the early Universe should be detectable as variations in the temperature of the cosmic microwave background radiation. However, measurements by Penzias and Wilson and others failed to detect fluctuations in the microwave temperature. Then, in 1992, the National Aeronautics and Space Administration (NASA) launched the Cosmic Background Explorer (COBE) satellite, which was built to detect such variations, and did, indeed, find that the microwave background contains tiny temperature fluctuations. This discovery was considered to be further support for models of the Universe based on the Big Bang theory. Again, a prediction made by theory was verified experimentally.

It has been proposed that there was an initial phase of extremely rapid expansion of the early Cosmos, followed by a slower rate of growth in size. In this 'inflation' period, the Universe expanded from one ten-thousand-trillion-trillionth of a centimetre to 100 centimetres (3 feet) across in the initial fraction of a second.

The density of matter in the Universe is critical for the fate of the Cosmos. If the mass density is above a critical level, the Universe will eventually collapse back upon itself in a 'Big Crunch'. If the density is below that critical level, the Universe will expand forever. If, however, the density is exactly at the critical level, the Universe will continue to expand, but at a gradually decreasing rate. Some physicists believe that the mass density of the Universe is at the critical level that will allow it to expand at an increasingly slow rate. In this state, it has been called a 'flat' universe. However, if this is correct, or indeed if the mass density is greater than the critical level, it means that we have not yet detected more than nine-tenths of the matter that exists in the Universe. It may be that we are surrounded by this 'missing' matter. If that is the case – and nobody knows for sure whether it is or not – it must be composed of particles that can pass through the ordinary matter with which we are familiar. Sophisticated underground equipment has been made that might detect this so-called **dark matter**. If it is found, it will be one of the greatest discoveries in physics, and it will mean that the matter we know now is just a small proportion of that which exists.

There are still many questions to be answered about the early stages of the Universe's existence. Before the Planck time, the General Theory of Relativity breaks down and a new theory that deals with the very small is needed to understand this phase. Physicists are using quantum theory (Chapter 5) to try and work out what the Universe was like in this tiny fraction of a second after the Big Bang. The new science of quantum cosmology, the study of the first one ten-million-trillion-trillion-trillionth of a second after the Big Bang, has arisen out of these considerations. In particular, a quantum theory of gravity is needed. Whereas the other three fundamental forces of Nature – weak nuclear interactions, electromagnetism and strong nuclear interactions – have been described by quantum theory, gravity remains elusive in this respect. Some theories, for example that of 'superstrings', have been developed to encompass all four forces in terms of quantum theory, but none of them has yet given a conclusive description of the Universe before the Planck time. Perhaps the detection of 'gravity waves' will provide some clues. These waves are predicted to exist, but nobody has yet found them, although several groups of physicists are searching for them. Their signature may yield information about the initial stages of the Universe, just as the cosmic microwave background radiation was informative about the early Cosmos.

What of the singularity – the state of infinite density and infinite space–time curvature that Einstein's General Theory of Relativity predicts existed at the 'moment' of Creation? The British scientists, Stephen Hawking (b. 1942) and Roger Penrose (b. 1931), showed that singularities must exist if one applies General Relativity to the Universe; if one wishes to avoid them, new theories must be developed. Based on his studies of black holes, which are also predicted to have a singularity at their centre, Hawking developed a theory that incorporates General Relativity and quantum theory that does go some way to providing a model of the earliest moments of the Cosmos, and it avoids the singularity. This theory is called a 'theory of everything' like several others being developed, because it explains the Universe right from its origins to the present. It may be that the Universe did not have a beginning, that the idea of 'zero time' is meaningless. One such scenario is that the Universe simply 'appeared' from a state in which time has no meaning, but which can be described using quantum theory; after the Planck time, it evolved according to General Relativity. Some physicists believe that a correct 'theory of everything' is not far off. Then we shall have a mathematical description of the Universe from its 'creation' to the present.

8
Molecular soccerballs

In December 1991 the journal, *Science*, which is highly respected by top scientists around the world, announced the winner of its annual award of 'Molecule of the Year'. This title is given to the scientific development during the previous twelve months that is deemed to have exemplified quintessential scientific endeavour. The 1991 winner was a molecule called **buckminsterfullerene**, or C_{60}. The discovery of this elegant and remarkable molecule is a story of excitement, serendipity and inquisitiveness. It illustrates beautifully how the scientific process works and shows how even well established ideas can be modified by a new and unexpected discovery.

The discovery of buckminsterfullerene, which is a form of pure carbon, has unveiled a novel type of molecule together with a brand new field of chemistry. Before buckminsterfullerene was known, carbon was thought to exist in only two naturally occurring crystalline forms: diamond and graphite. Other, non-crystalline forms, such as soot, are also known. Many scientists were quite taken aback when, in 1985, buckminsterfullerene was discovered and found to be a third form of crystalline carbon. Since then, many other molecules related to buckminsterfullerene have been discovered or made: they all belong to a class of substances called **fullerenes**.

To many chemists, fullerenes are far more interesting and exciting than graphite or diamond. These substances could lead to the production of hundreds, even thousands, of novel substances with chemical and physical properties never before seen. The future of fullerenes could well involve the manufacture of a variety of substances, including new synthetic polymers, industrial lubricants, superconductors, molecular computers and medically useful drugs.

The discovery of buckminsterfullerene is reminiscent of the proposal of the ring structure of benzene by the German chemist, Friedrich Kekule (1829–1896), in 1865. Chemists at the time knew how many carbon and hydrogen atoms a benzene molecule had (six of each), but they were at a loss to explain how these atoms were arranged. Faraday had first discovered benzene in 1825, and the German chemist, Johann Loschmidt (1821–1895), later suggested that the benzene molecule was cyclical. However, it was Kekule who finally proposed the proper structure of benzene. Kekule reputedly had a dream in which he saw a string of six carbon atoms turn into a snake. The snake bit its tail, forming a ring. Soon afterwards, Kekule proposed the six-carbon ring structure of benzene, opening up a whole new area of chemistry ('aromatic chemistry') that led to many of today's synthetic substances, from dyes to drugs. If Kekule's revelation is considered to be the beginning of the chemistry of ring compounds, buckminsterfullerene can be said to herald the dawn of 'sphere' chemistry.

Fullerene chemistry has it roots in research that was being carried out in the field of astronomy. The scientists involved in the discovery of buckminsterfullerene were initially interested in the chemicals present in the dust that surrounds and occurs in between stars. If the diverse benefits that are predicted to arise from fullerenes are forthcoming, and there is every reason to be confident that they will be, then buckminsterfullerene will be a supreme example of how basic research into the stars of the Cosmos can produce medically and industrially valuable substances.

In order to appreciate the context in which buckminsterfullerene was discovered, it is useful to have some understanding of the molecular structure of the two other crystalline forms of carbon, namely diamond and graphite.

Diamond, graphite and carbon stars

Diamonds have been mined for thousands of years and were known even in prehistoric times. Although they are made of the same basic element (carbon) as soot, there is no doubt as to which of these substances is the most aesthetically pleasing to the eye at the macroscopic level: diamonds, but not soot particles, are traditionally said to be a girl's best friend. Diamond is the hardest known natural material. Its hardness contributes not only to its robustness as a jewel, but also allows it to serve a very useful purpose as an industrial cutting tool and grinding agent. Graphite, familiar to all of us as the 'lead' in pencils, contrasts distinctly in appearance and properties with diamond: it is darker in colour and much softer. Its flakiness allows graphite pencils to be sharpened easily. Graphite is also useful in industrial machinery, and in the electrical and space industries, particularly because it acts as a lubricant and electrical conductor.

Why are diamond and graphite, which are both essentially pure carbon, so very different from each other? The answer lies in the arrangement of carbon atoms within their molecular structures. In diamond, each carbon atom has four strong chemical linkages joining it to four other carbon atoms in the shape of a tetrahedron (Figure 15). Each of these four carbon atoms is, likewise, joined by four strong bonds to another four carbon atoms, and so on, to produce an enormous lattice of cross-linked carbon atoms held rigidly in place. This arrangement strengthens the structure – the links between the carbon atoms are extensive and difficult to break or deform. Diamond is consequently very hard indeed.

Graphite, on the other hand, has a completely different structure. Here, each carbon atom is joined to only three other carbon atoms by strong linkages. This produces an extended structure consisting of sheets of hexagonal rings of carbon atoms (Figure 15). Graphite consists of many layers of these sheets of carbon hexagons arranged one on top of the other. Although each sheet is rigid, there are only weak interactions between adjacent sheets, allowing them to slide over each other. It is this property of graphite – the ability of layers of carbon atoms to move over each other – that provides it with its flakiness. Graphite also conducts electricity well, whereas diamond does not. The reason for this is that electrons are free to move around within the sheets of carbon hexagons in graphite, whilst all of the electrons of diamond are used in forming the strong links between carbon atoms and are not free to move around.

In the early 1980s carbon was thought to exist in only two crystalline forms, diamond and graphite. Various scientists around the world were, however, interested in the carbon that occurs in space, particularly in the materials around stars. In the 1970s, black clouds of interstellar dust that stretched across our own Milky Way galaxy were shown to contain molecules having short chains of carbon atoms. Some scientists thought that these clouds were being produced by the **red giants** (carbon stars). Red giants are formed from dying stars that have distended and started to cool down as a result of exhaustion of the nuclear energy that is the source of starlight. Such stars are visible in the night sky: for instance Betelgeuse in Orion is a red giant. They frequently emit large quantities of dust, and one theory was that this dust contained particles of carbon, perhaps resembling soot.

Professor Harry Kroto (b. 1939) and his team at Sussex University, England, were one group of scientists who were particularly interested in trying to determine the structures of the carbon molecules produced by red giants. Kroto and his collaborators had identified several carbon-containing molecules in interstellar dust in the 1970s, and they were on the lookout for even longer molecules. Meanwhile, a collaborative effort was under way between Professor Don Huffman (b. 1935) at the University of Arizona, USA, and Professor Wolfgang Kratschmer (b. 1937)

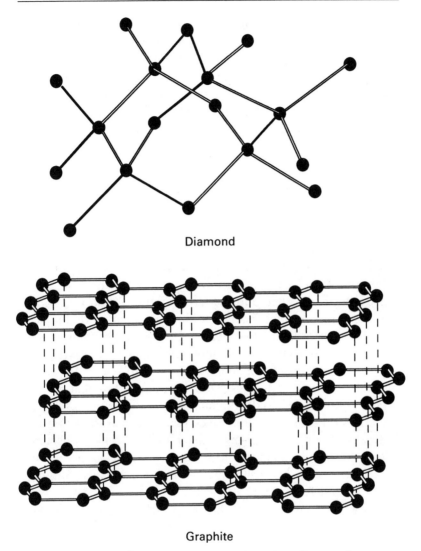

Diamond

Graphite

Figure 15. Structure of diamond and graphite. Diamond contains carbon atoms each linked to four other carbon atoms, to produce an extensive rigid structure. In graphite each carbon atom is strongly linked to three other carbon atoms, producing layers of hexagonal rings. Carbon atoms are shown as black circles; strong linkages are solid lines; weak interactions are depicted as dotted lines.

at the Max Planck Institute for Nuclear Physics in Heidelberg, Germany, to try and produce dust similar to that in space. Huffman and Kratschmer believed that this interstellar dust consisted mainly of carbon, so, in order to simulate it, they vaporised graphite by passing an electric current through two graphite rods enclosed in a vacuum. The vaporised graphite produced a cloud of black smoke, which the scientists examined closely.

Discovery of buckminsterfullerene

In 1982 Huffman and Kratschmer examined their simulated interstellar carbon dust using a method in which ultraviolet light is passed through the dust. Information can be obtained on the nature of the molecules present from the way in which the ultraviolet light is absorbed and scattered by the carbon dust. They compared the results obtained with their 'soot' with the data obtained with ordinary soot obtained simply by burning coal in the normal atmosphere of the laboratory. A surprising difference between the two forms of soot struck the two scientists: several particular carbon molecules were especially abundant in soot obtained from vaporised graphite. These molecules gave particularly strong blips on the traces that recorded the ultraviolet light absorption experiments, and Huffman and Kratschmer called the result obtained with the vaporised graphite their 'camel spectrum', because of its two prominent blips. Although Huffman and Kratschmer did wonder what the two peaks were and why they stood out above the other carbon molecules in their carbon dust, they decided not to pursue the work any further, at least for the time being.

Huffman and Kratschmer's 'camel spectrum' remained essentially unstudied for five years. When they returned to it, buckminsterfullerene – the main cause of the 'hump' – had already been discovered by Harry Kroto, in collaboration with Bob Curl (b. 1933), Richard Smalley (b. 1943) and their colleagues at Rice University in Houston, Texas.

Kroto visited Curl and Smalley at Rice University in 1984.

The three scientists agreed that they should collaborate in an effort to learn more about the carbon molecules in interstellar space. They began experimenting using a device at Rice University in which graphite was vaporised at 10 000 degrees Centigrade using a powerful laser beam. This device, as with Huffman and Kratschmer's apparatus, might, Kroto thought, simulate the conditions around red giants. The 'soot' produced by this method was examined for its molecular composition. It was found to contain carbon molecules having between thirty and one hundred carbon atoms. Interestingly, one of these carbon compounds was much more abundant in the soot than any other compound: it was a molecule containing sixty carbon atoms and was therefore called C_{60}. Another molecule containing seventy carbon atoms (C_{70}) was also relatively abundant. This meant that C_{60} and C_{70} were particularly stable molecules in the soot obtained from vaporised graphite. Why was a molecule containing sixty carbon atoms so much more stable than the other molecules?

Kroto, Curl, Smalley and their co-workers were perplexed by their result. C_{60} could not be ignored: it came up time after time in their experiments and was something to be pursued, if only out of sheer curiosity. They started to consider how sixty carbon atoms could be arranged to produce a stable molecule. One idea that their discussions led to was that the layers of graphite hexagons had wrapped themselves up into closed cages. Kroto remembered the geodesic dome at EXPO '67 in Montreal. This dome, one of thousands of geodesic domes erected around the world, was designed by the US architect, Richard Buckminster Fuller (1895–1983). These domes are composed of hexagons and pentagons. Kroto wondered if C_{60} might be a kind of molecular geodesic dome. Kroto also remembered a three-dimensional map of the stars he had made some years before for his children: it contained hexagons and pentagons. Could C_{60} be made up of carbon atoms arranged in the form of pentagons and hexagons?

Smalley cut out pieces of paper in the shape of hexagons and pentagons and pieced them crudely into a three-dimensional spheroid form that had sixty corners, each one supposedly corresponding to a carbon atom. He used twelve pentagons and twenty hexagons. Smalley did not at first realise what the exact structure he had made was. He called the mathematics department at Rice University and asked them if they knew what structure he had formed. The answer that came back was simple: the model Smalley had made was identical to the shape of a modern soccerball (Figure 16). The modern soccerball is made up from twelve pentagonal patches (usually black) and twenty hexagonal patches (usually white) sewn together into a sphere. The sixty carbon atom molecule in soot obtained from vaporised graphite was a molecule-sized soccerball!

Kroto, Curl, Smalley and co-workers agreed that C_{60} should be named after Buckminster Fuller, whose geodesic domes had entered significantly into their thoughts: it should be called buckminsterfullerene.

Unfortunately, buckminsterfullerene was available only in small quantities and it was necessary to obtain much more of it in order to carry out studies designed to confirm its soccerball shape. At this stage, Huffman and Kratschmer conjectured that the 'hump' on their 'camel spectrum' might be buckminsterfullerene. After all, the 'hump' was produced when soot from vaporised graphite was examined. Huffman, Kratschmer and their colleagues confirmed this suspicion and went on to be the first scientists to extract relatively large amounts of buckminsterfullerene and to produce crystals. Buckminsterfullerene joined diamond and graphite to become the third known form of crystalline carbon.

This amazing molecular soccerball caused a sensation amongst chemists and physicists. Scientists in laboratories all over the world began working on buckminsterfullerene and its fullerene relatives. Its structure has been confirmed since its isolation in large quantities in 1990. It has also been found in ordinary soot from a bunsen burner, which means that we have all made buckminsterfullerene in our school chemistry lessons.

The other especially stable molecule produced in vaporised graphite soot that was shown to contain seventy carbon atoms (C_{70}) has been shown to be egg shaped: it resembles an elongated sphere, perhaps a little like a rugby ball. Like buckminsterfullerene, C_{70} also consists of twelve pentagons, but it has twenty-five hexagons.

There is a mathematical principle which states that twelve

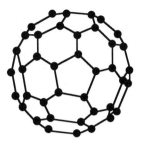

C$_{60}$ (buckminsterfullerene)

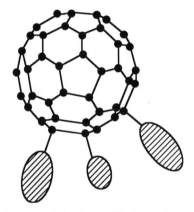

Figure 16. Structure of buckminsterfullerence (C$_{60}$) and some possible derivatives of it. Sixty carbon atoms are arranged in the form of a sphere consisting of 12 pentagons and 20 hexagons (top). Atoms and other molecules (large circle) can be incorporated inside the sphere (bottom left), and various chemical groups (shaded ovals) can be attached to the sphere (bottom right). This provides C$_{60}$ with tremendous versatility, promising applications in areas from medicine to electronics. Carbon atoms are shown as black circles.

pentagons are required for complete closure of a cage consisting of pentagons and hexagons. Hexagons alone, however many there are, can never produce a closed spheroid. In fact, any number (except one) of hexagons can be made into a closed spheroid if there are twelve pentagons associated with them. The smallest closed structure that can be built requires two hexagons and twelve pentagons: this would be equivalent to C_{24}. Chemical stability of structure is particularly good if none of the twelve pentagons has any of its five sides in common with any side of another pentagon. In other words, if the pentagons are isolated from each other by hexagons, the spheroids so formed are especially stable. Critically, C_{60} (buckminsterfullerene) is the smallest closed structure that can be made in which twelve pentagons are isolated from each other by hexagons, and this is what makes it so special. As the number of carbon atoms is increased from sixty, the next closed structure that can be formed in which those twelve pentagons are isolated completely by hexagons is C_{70}. Molecules containing between sixty and seventy carbon atoms were not as prominent in the experiments on graphite soot as C_{60} and C_{70} because all of these intermediate molecules would have pentagons adjacent to each other and would therefore not be as stable.

A host of fullerenes has been produced since the discovery of buckminsterfullerene. Closed spheroids containing as many as five hundred and forty carbon atoms (C_{540}) have been detected, and incompletely closed molecules called 'buckytubes' have been produced. Buckminsterfullerene has been nicknamed the 'buckyball'. Chemical compounds of fullerenes with other molecules are proving to be particularly interesting and important. Molecules can be made, for example, in which metal atoms and other chemicals are 'trapped' inside the soccerball of buckminsterfullerene. Some of these substances may have novel properties such as superconductivity. It may be possible to introduce chemicals, such as drugs, inside the soccerball and to create 'doors' on the surface of the sphere that open when the drug needs to be released, that is at the right time and place in the body. Other substances have been made in which chemical groups are added to the outside of the soccerball molecule, creating 'ears' (one such derivative has been given the name, 'bunnyball'). These substances are a gold-mine of new chemicals and only time is needed before new and exciting applications of this soccerball chemistry are forthcoming.

It is still not clear whether or not buckminsterfullerene does occur in the interstellar dust spewed out by red giant stars. What is clear is that man's quest to understand the stars has led serendipitously to a brand new and exciting area of science whose applications to society may yet turn out to be enormous. As Harry Kroto said of the buckminsterfullerene discovery, 'This advance is an achievement of fundamental science, and serves as a timely reminder that fundamental science can achieve results of importance for strategic and applied areas.'

9
Jostling plates, volcanoes and earthquakes

W hilst astronomy and cosmology have produced some truly
remarkable and wonderful discoveries, studies of the
structure and dynamics of our own planet Earth have also come
up with some scientific surprises. One of the great developments
of geophysics (the study of the physics of the Earth) was **plate
tectonic theory**, which was proposed in the 1960s to account for
many of the observations regarding the structure and properties of
continents and oceans, and which was of very great significance
for our understanding of the formation of mountains and the
occurrence of volcanoes and earthquakes. It also provided valuable insights into many other processes that occur on Earth,
including the evolution of life, climate changes and the structure
and properties of the ocean floor. Plate tectonics completely

revised the way in which we see our planet. It can be considered to be the geophysical equivalent of the Big Bang theory, which was proposed to explain the origins of our more extended home, the Universe (Chapter 7).

Most of us take the Earth so much for granted that we rarely, if ever, stop to ponder the reasons for the existence of oceans and continents. Although we view volcanoes, mountains and earthquakes with awe and sometimes trepidation and fear, only infrequently do we consider the mechanisms by which these great features of the Earth came to be. Most of us have some knowledge of the history of humanity or of the evolution of living creatures, but how many of us can say that we have even a rudimentary picture in our mind's eye of the appearance of the Earth's surface between the time when the planet was formed, more than four and a half billion years ago, and the present?

Volcanic eruptions and earthquakes have occurred throughout history and not a year goes by without one or both happening somewhere in the world. Often the event is relatively harmless, but every now and again a volcano or an earthquake causes a tragic human disaster. The ancient city of Pompeii in Italy is a fine example of the catastrophic effects that volcanic eruptions can have on human society. On 24 August AD 79, nearby Mount Vesuvius violently erupted, leaving Pompeii buried under a layer of ash six metres (twenty feet) deep. Within the space of a day, the city had ceased to exist. Human bodies disintegrated when they came into contact with the molten lava, leaving their imprints, which were preserved when the lava cooled and solidified around the bodies. Centuries later, archaeologists made casts of the unfortunate victims who had met with such a fate, by pouring cement into the human hollows formed in the solid volcanic rock.

Krakatoa, an island between Java and Sumatra, was the site of one of the most devastating volcanic eruptions ever recorded. In 1883, the volcano erupted intermittently over a period of several months; one of these eruptions was so loud that it was heard 5000 kilometres (3000 miles) away and was equivalent in its explosive power to several thousand atom bombs. Nearly forty thousand people were killed as a result of Krakatoa's eruption, the majority of the deaths being due to the enormous tidal waves created by the volcanic explosions. Half of Krakatoa was blown away during the eruptions. More recently, in 1980, Mount St Helens in Washington State, USA, erupted, blowing away the top third of its peak. Vast numbers of trees were felled and stripped of their bark and damage was estimated at billions of dollars.

Earthquakes have also produced their share of destruction and human misery. Nearly a million people died in an earthquake that occurred in China in 1556, and more than half a million people are thought to have been killed in another Chinese earthquake in 1976. San Francisco, California, suffered a well known earthquake at twelve minutes past five on the afternoon of 18 April, 1906. The city's ground was shaken and torn apart and the downtown area was almost completely destroyed; hundreds of thousands of people lost their homes and nearly eight hundred people were killed. Plate tectonic theory has allowed predictions of the future state of the Earth to be made. For California the prognosis is that the western regions of the state, including Los Angeles and parts of San Francisco, will, within the next twenty million years, become a small continent separated from the American continent.

Our ideas about the origin and properties of volcanoes and earthquakes were altered dramatically during the second half of the twentieth century as a result of plate tectonic theory. The present state of the Earth is a consequence of its past. With the improved predictability of the future condition of the planet, human devastation caused by earthquakes and volcanoes might be minimised in the years to come. Buildings and other structures can be built with a greater ability to withstand these events and people can become better prepared to evacuate their towns and can be given sufficient warning to do so.

Plate tectonics incorporates and extends an earlier idea proposed in 1912 by the German climatologist, Alfred Wegener (1880–1930), who believed that the present-day continents were, about two hundred million years ago, joined together into one large supercontinent, which he called **Pangaea**. According to Wegener's theory, this supercontinent eventually broke up and its fragments drifted around the surface of the Earth to produce today's continents. Unfortunately, Wegener's ideas were un-acceptable to most other scientists of the day and were received with tremendous ridicule amongst many geophysicists.

Before Wegener's ideas about continental drift are discussed, it is useful to have some idea about the structure and composition of the Earth.

Structure of the Earth

Basically, the Earth consists of a solid, central inner **core**, which is about 1300 kilometres (800 miles) thick and consists mostly of iron. Above this inner core there is a liquid iron outer core, which is a further 2200 kilometres (1400 miles) thick. The solid and liquid iron cores together occupy about a third of the total mass of the Earth and a sixth of its volume. Further towards the sur-face of the earth, just above the iron core, there is the **mantle**, which is nearly 3000 kilometres (1900 miles) thick and consti-tutes two-thirds of the mass of the Earth and more than four-fifths of its volume. Finally there is the Earth's **crust**, which rests on the mantle and consists of the ocean floors and the continents. The crust comprises less than a half of one per cent of the mass and seven-tenths of a per cent of the volume of the Earth.

The land that we see above sea-level represents a small fraction of the depth of the Earth's crust and therefore an even tinier proportion of the Earth's substance. The rocks that make up the Himalayas, for example, form crust that is 70 kilometres (45 miles) deep, but we see only the tips of the mountains: Mount Everest, the world's tallest mountain, is less than 9 kilometres (5.6 miles) above sea-level, so most of it is buried below ground.

The dividing line between the Earth's mantle and the overlying oceanic and continental crust is called the **Moho**, or the **Moho-rovicic Discontinuity**, in honour of the Yugoslavian scientist, Andrija Mohorovicic (1857–1936), who discovered it. The conti-nents and ocean floors float on the mantle, which is made up of denser materials than the crust. Ocean floors are made up of basalt and other dense rock, and they are relatively thin. Continents, on the other hand, are composed of less dense granite rocks and are thicker than the ocean floors. This gives ocean floors a lower buoyancy than continents, and this explains why the Earth's crust is essentially a two-tiered structure, with the ocean floor below the continental land. The oceans cover about three-quarters of the surface of the Earth, the rest being the continents. Continents are, on average, 30–40 kilometres (19 to 25 miles) thick and ocean floors are 6 to 7 kilometres (3.7 to 4.3 miles) thick.

Wegener knew a good deal about the composition of the Earth's crust. In particular, he knew that the ocean floor is less buoyant than the continental crust, and that the continents are always elevated above the ocean floor. It was this characteristic, along with other evidence, that led him to propose his theory that continents drifted.

Wegener's theory of continental drift

Wegener amassed a great deal of evidence for his theory of **continental drift**, which should have been considered by objective thinkers to be at least plausible. Its rejection may have been because it challenged conventional wisdom that the Earth was static: this belief was deeply ingrained in the minds of the world's intellectuals. How could enormous continents move around the Earth's surface as if they were rafts on a river?

Wegener was initially compelled to consider the idea of continental drift as a result of a simple observation that he and others before him had made: that the shorelines of the coasts of eastern South America and western Africa appear to fit together like two pieces of jigsaw puzzle (Figure 17). We know today that these two continents fit into each other even more tightly if the shapes of their coastlines are examined at a depth of a kilometre below sea level. Wegener noticed similar matches between the shorelines of other continents and this led him to postulate the existence of the supercontinent Pangaea and to propose its break-up and drift to give rise to today's continents. Theories that were previously propounded to explain this match included the idea that Africa and South America were once part of the same land but became separated by water (the Atlantic Ocean) when the land joining the two continents sank underneath the ocean. Some nineteenth century scientists postulated that the floor of the Atlantic Ocean contained the remains of a lost continent, Atlantis. Wegener argued against this theory because the ocean floor was known to be less buoyant than the continental crust which consequently meant that the ocean floor always had to be below the level of the continents. It was more logical, Wegener said, to assume that the two continents had drifted apart and that new ocean floor had formed in between them. There was strong evidence that continental land was able to move slowly upwards. For example, centuries-old moorings for boats on some harbour walls appeared to have risen so high in some regions of the world that boats could no longer be attached to them. If land could rise, Wegener asked, why could it not also move from side to side?

Wegener produced more evidence for continental drift. The plants and animals around the shores of eastern South America and the west coast of Africa were similar to each other, and this was particularly true of fossilised life forms. This would make sense: at some stage in the Earth's past, when the continents were joined, the now-fossilised organisms would have coexisted on the same land and so would resemble each other very closely. Later on, after the continents had drifted apart, there would have been some evolution of life, and species would still be similar on the two continents but they would be somewhat more diverse than the fossilised creatures that once lived together on the same piece of land.

Other evidence for continental drift was that mountain ranges on different continents whose coastlines matched also fitted together nicely. Moreover, the layers of different rocks on the two continents that were said to have been joined in Pangaea were very similar, and geological studies predicted that the two continents once shared similar climates. Unfortunately, Wegener

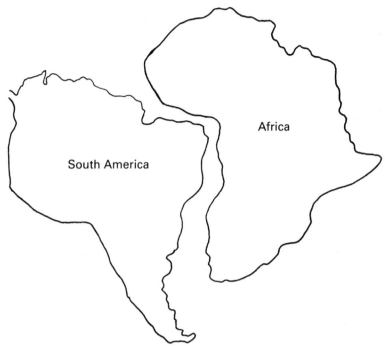

Figure 17. Match between the eastern coastline of South America and the western coastline of Africa. The two continents appear to fit together like pieces of a jigsaw puzzle. This, and similar matches between several other continents, inspired Alfred Wegener to consider the possibility that the continents were once joined together to form a supercontinent (Pangaea).

had a problem with explaining the precise mechanism by which continental drift occurs, and this is probably another reason why his opponents so vehemently disagreed with him. When Wegener died in 1930, few were convinced of his theory of continental drift: no amount of evidence was enough to shake off established ideas regarding the static nature of the world's lands. Wegener's proposal of continental drift remained an unsung scientific breakthrough for three more decades. Then, in the 1960s it was revived and incorporated into plate tectonic theory.

Plate tectonics

Several developments took place between Wegener's death and
the appearance of plate tectonic theory. First, studies of the magnetic properties of continental rocks revealed some peculiar features that were best explained if continents had, indeed, drifted
in the past. Second, after the Second World War, improved technology allowed the ocean floors to be examined in a greater depth
than ever before. This came about partly as a result of increased
military awareness that it was important to understand the oceans,
especially to allow submarines carrying nuclear bombs to travel
at great depths in the oceans. The new knowledge of the ocean
floors revealed some interesting features that could be explained
best if they were incorporated into a theory involving moving
continents.

The magnetic data that helped to revive the idea of continental drift was based on the principle that many rocks retain a
'memory' of their past orientation with respect to the Earth's
magnetic poles. This is especially true for volcanic rocks. The
atoms of iron and other magnetic substances in molten lava align
themselves in a magnetic north–south orientation just as
compass needles point in the same direction. When such lava
solidifies after it has cooled down, the atoms of magnetic substances in it remain locked in the same orientation that they
had when they were molten. Therefore, the direction in
which such magnetic components of today's solid rocks are
oriented reveals the direction of the Earth's magnetic poles
at the time those rocks solidified. The extent to which such
'fossilised compass needles' dip also reveals the distance of
the rocks from the Earth's magnetic poles at the time the rocks
were molten. At the poles, the magnetic particles point downwards vertically; as they approach the equator they dip less; at
the equator they do not dip at all. In other words, from measurements of the magnetic properties of rocks, one can deduce the
direction of the poles as well as the latitude that the rocks occupied at the time they solidified. If continents did drift, magnetic
studies might reveal this motion, since the continents would

presumably have changed their position of latitude during their migration.

In the 1950s, the British geologist, Stanley Runcorn (b. 1922), and his colleagues, obtained some peculiar results when they examined the magnetic features of rocks in Europe. When magnetic properties were investigated in rocks of different ages from the same geographical area, the Earth's magnetic poles appeared to have changed position at different times during the Earth's past. There were two major possible interpretations of this research: either the poles had drifted and moved around the globe during the Earth's geological history, or Europe had drifted and the poles had remained in the same place. Most scientists remained glued to the static model of the continents and either rejected the data as being unreliable or accepted the idea that the magnetic poles, but not the continents, had migrated.

When similar magnetic properties of rocks were examined in other continents, the poles there were also found to have had apparently different positions at different times in the Earth's past. When the predicted paths of migration of the poles as seen from different continents were plotted on the surface of the Earth, however, they did not coincide. If there was one pole, it should obviously have occurred at the same position at the same time for all continents. If the continents did not move, the only other explanation now was that there were several magnetic poles in the Earth's past, which was thought extremely unlikely. However, if the continents were considered to have been juxtaposed at some time in the past, in a manner similar to that proposed for Pangaea by Wegener, then the plotted paths of polar migration did coincide for different continents. This agreed excellently with the idea that the continents had, indeed, been joined together and had later drifted apart. The magnetic poles were static, according to this interpretation, and the continents moved. More scientists were now convinced that continental drift was a possibility, although the idea was still difficult for many to accept.

When detailed knowledge about the ocean floors became available, continental drift was finally accepted, and plate tectonics was born. Much of this knowledge of the ocean floor came from research carried out in the USA by Bruce Heezen (1924–1977),

Henry Menard (1920–1986) and their co-workers. The ocean floor was revealed as a realm where mountains and trenches were more magnificent and awesome than they were on land. One prominent feature of the ocean floor is the **mid-ocean ridge system**, which meanders around the globe. This ridge system consists of rugged structures that extend for a distance of 60 000 kilometres (37 500 miles); they are several hundred kilometres/miles wide and more than 4500 metres (15 000 feet) high. The mid-ocean ridge system is the site of frequent earthquakes and volcanic activity. Most of the volcanic activity occurs on the ocean floor: a rift valley containing molten lava runs along the ridge's length. In some places, for example Iceland, the volcanoes occur above sea-level. The ridge is not completely continuous: at various intervals along its length ridges and troughs, called **fracture zones**, occur at right angles to it, creating what appears to be a stitch-like pattern. The San Andreas fault in California is another example of a fracture zone.

Another outstanding feature of ocean floors is the presence of **trenches**, which are above five kilometres (three miles) deep, and, like the ridge system, are sites of earthquakes and volcanoes. The deepest parts of the oceans occur where trenches exist. Trenches are often associated with chains of islands, for example those associated with the Aleutian Islands in the Pacific Ocean.

Several theories were proposed to explain these characteristics of the ocean floor's geophysics. An early idea was that the Earth was expanding, causing the sea floor to crack; but there was no evidence for this theory. Another theory, which was widely accepted, was proposed by the US scientist, Harry H. Hess (1906–1969), and was called the 'sea floor spreading' hypothesis. The sea floor spreading hypothesis was an extension of an earlier idea that the mid-ocean ridge formed as a result of heating of the Earth's crust, which, in turn, was said to be due to the underlying mantle being heated by the high temperature of the Earth's core. According to Hess, the Earth's crust 'floats' on the flowing mantle. As the mantle material heats up, it rises and heats up the crust; this causes the crust to expand and crack in vulnerable regions. The mantle material rises through the crack and oozes out onto the ocean floor as molten lava. As it flows upwards through the crack, the lava pushes the crust on either side of the crack outwards. The lava flows out of each side of the crack and cools down as it moves away from the centre of the crack. During this cooling, it solidifies and forms a new layer on top of the existing ocean crust.

According to Hess, the Earth was not expanding, and this meant that the new crust formed at the cracks, which correspond with the crests of the mid-ocean ridge, had to be compensated for by a loss of existing crust elsewhere. Hess proposed that the trenches that occurred on the sea floor were the sites of destruction of sea floor crust: the expansion of the sea floor at ocean ridges caused sea floor crust to be forced into the mantle at trenches. This theory explained why it was that rocks obtained from the sea floor were always found to be less than a quarter of a million years old, whereas continental rock had been found that was more than three and a half billion years old. The ocean floor was made of young rocks, since it was continually being replaced by newly solidified lava from the underlying mantle; and old ocean crust was, at the same time, being forced downwards into the mantle at trenches, there to become molten.

Hess's theory also required that the rocks of the ocean floor should be older the further away they are from the centre of the mid-ocean ridge, since new lava starts in this region and, over millions of years, slowly spreads outwards. In other words, the newest rocks should occur at the ridge crests and the oldest at the trenches. Hess's theory also provided the first acceptable mechanism for continental drift: as ocean floors were renewed at ridges and destroyed at trenches they would carry the continents with them, rather like objects being carried on a conveyor belt.

Evidence for sea floor spreading was forthcoming, particularly when the age of sea floor rocks were determined from their magnetic properties. This allowed the age of rocks on either side of ridges and trenches to be obtained, and the data was in excellent agreement with the ideas proposed by Hess. Indeed, when these rates of renewal and spread of the sea floor were measured, they were very close to those expected from Wegener's idea of continental drift that was proposed decades earlier. This was a

clear indication that continental drift did occur and that it was
due at least partly to the cycles of renewal and destruction of the
ocean floor attached to the adjoining continents.

In the late 1960s the British geophysicists, Dan McKenzie
(b. 1942) and Robert Parker (b. 1942), and (independently) the
US geophysicist, Jason Morgan (b. 1935), brought together the
ideas of Wegener, Hess and other geophysicists who had contributed to our understanding of continental drift and knowledge
of the continents and ocean floor. They suggested that the Earth
contains a number of plates (there are about a dozen large plates
and several smaller ones) that consist of the continental and
oceanic crusts as well as the upper regions of the Earth's mantle.
In other words, these plates are more than just crust: they also
contain mantle material. The name given to the layer of the
Earth containing the plates is the **lithosphere**. Plates are about
100 kilometres (60 miles) thick and have, as their edges, the
mid-ocean ridges, the ocean floor trenches and fracture zones.
They float on an underlying layer of the mantle called the
asthenosphere and may contain either continental crust or ocean
floor crust or both. The continents and ocean floors are carried
by the floating plates like objects sitting on rafts. When tectonic
plates break up, old continents may give rise to several new ones,
and these eventually may drift apart, as was the case when
Pangaea broke up.

When two plates collide with each other, the outcome depends
on the particular features of the plates (Figure 18). If the edges
of two plates containing oceanic crust collide, either one of them
will descend under the other. When the continental edge of one
plate collides with the ocean crust edge of another, the oceanic
plate always descends under the continental one. The reason for
this is that continental crust is more buoyant than ocean floor
crust, and so continental crust always rises above the ocean
floor. When the edges of two colliding plates both consist of
continental crust, neither will sink because of their high buoyancy;
instead, the boundary may become crumpled and produce mountain ranges. The Himalayas were formed as a result of such an
encounter. Collisions between two continents may unite them
together into a single, larger continent.

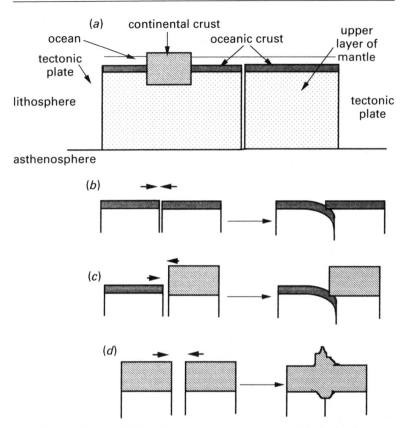

Figure 18.   Collisions between tectonic plates. (*a*) Each plate
consists of a region of the lithosphere (crust + upper part of the
mantle) and floats on the underlying mantle (asthenosphere).
Plates may contain either oceanic or continental crust, or both.
(*b*) When both edges of each colliding plate contain ocean crust,
one edge descends below the other. (*c*) When one edge contains
oceanic crust and the other continental crust, the less buoyant
oceanic edge descends below the continental edge. (*d*) When
both edges are continental, mountains arise, since continental
crust is too buoyant to be able to descend.

When two plates drift apart, new molten lava from the underlying mantle oozes up to fill the gap: this is the situation that
occurs in the mid-ocean ridge. At the trenches, ocean floor crust
from the descending side of a plate is destroyed and enters the

asthenosphere, there to become molten and eventually to be recycled as new ocean crust. The ocean floor is therefore constantly renewing itself as a result of plates colliding and moving apart. When continents crack, molten lava from the asthenosphere rises to fill the crack. Since this lava consists of basalt, it forms ocean crust when it cools down, and this means that new ocean floor is formed where the crack occurred. This allows a continent to split into two new continents separated by an ocean.

The net rate of increase of ocean floor material is essentially zero, because whenever new ocean floor is produced, an equal amount of old ocean floor plunges into the mantle and is destroyed elsewhere. However, this does not mean to say that new ocean floor formed at a ridge is compensated for by loss of ocean floor from the same plates whose borders form this ridge. As long as old ocean floor is destroyed at the same rate that new ocean floor is made, the recycling processes can occur at borders of other plates. For example, the Mid-Atlantic Ridge system, which borders plates carrying North and South America, does not have its own trenches. When new ocean crust is made at this ridge, it is compensated for by a loss of crust from the Pacific plates, which border the west coast of the Americas. In this way, the plates containing the Americas are growing larger in area at the expense of the Pacific plates.

Plates may also slide past each other instead of colliding with one another or moving apart; when this occurs, earthquakes frequently result. Fracture zones occur where plates slide past one another. Earthquakes also occur in other places where movement of the Earth's lithospheric plates is intense; they are particularly common at mid-ocean ridges, where two plates move apart, and at trenches, where one plate descends under another. If the major earthquakes that have been recorded are plotted on a map of the Earth, they correspond strikingly to the edges of plates (Figure 19). Similarly, volcanoes map very closely with plate boundaries. Volcanoes do not occur where two plates collide (to produce mountains); nor do they occur where continents slide past each other. They do occur where plates move apart or where one plate descends under another following a collision. When an oceanic edge of a plate descends under another plate, some of the ocean

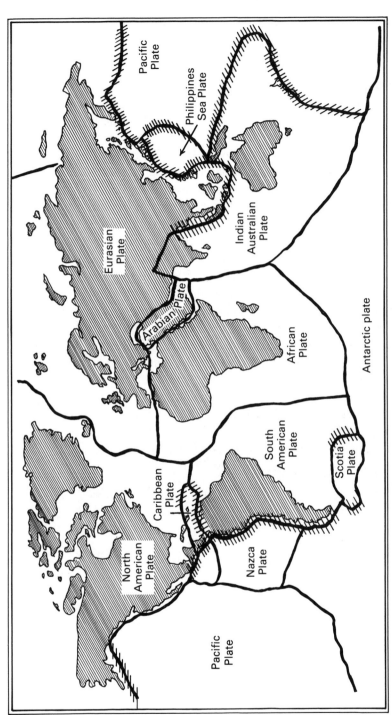

Figure 19. Volcanoes and earthquakes usually occur at or near the edges of tectonic plates. Plate boundaries are shown here as thick solid lines, and trenches are shown by lines crossing them.

crust and ocean sediment melt in the mantle and the molten substances have a tendency to rise to the surface and create active volcanoes. Japan and the Aleutian Islands contain numerous volcanoes that occur along this kind of plate boundary. Volcanoes also form where plates move apart and molten lava rises to the surface; examples include those along the eastern side of the Pacific plates and those in Iceland.

There is a good deal of evidence for the existence of tectonic plates and the consequences that arise from their collisions, separations and sliding can be seen clearly in many places on Earth. Tectonic plates are undergoing these processes today, and they will continue to do so in the future. Continents will continue to drift. Indeed, it has been suggested that all of the continents will, in several hundred million years' time, once more join together: the supercontinent so formed has already been given a name – **Neopangaea** – by some geophysicists. If Neopangaea does form it will undoubtedly break up again and the resulting fragments will drift apart to produce a geography that will be very different from the one we have today.

Human beings have created artificial, political and physical boundaries between their nations, but the natural processes that cause continents to break up and their fragments to drift apart and perhaps to rejoin over hundreds of millions of years are the real creators of the Earth's boundaries.

10
Soda water, phlogiston and Lavoisier's oxygen

P late tectonic theory explained some most important aspects
of the Earth's geophysical features. It also allowed the
evolution of life to be better understood in the context of the
state of the Earth in geological history. The continental land and
oceans, after all, provide the habitats of almost all plants, animals
and microbes, and so the dynamics of plate movement and its
effects, such as volcanoes, earthquakes, mountain formation,
ocean renewal and destruction, clearly will have had a tremendous
influence on the evolution of life on Earth.

The evolution of life into today's diverse species also depended
upon the appearance of oxygen in the atmosphere. Oxygen
accounts for a fifth of the volume of the air that we breathe.
Without it we cannot convert the energy of our food into the

metabolic energy required for muscular activity, growth and other bodily functions. In combination with hydrogen, oxygen produces water, H_2O, the medium in which life on Earth probably evolved and upon which it depends. Oxygen, alone or in combination with other elements, is the most abundant element in the Earth's crust and it is the fourth most abundant element in the Cosmos, behind hydrogen, helium and neon. In the upper layers of the atmosphere, oxygen gets converted to ozone by the Sun's ultraviolet rays. This ozone forms a blanket that protects life on the Earth's surface because it absorbs dangerous ultraviolet light. Knowledge of oxygen and what factors determine its rate of conversion to ozone is fundamental and essential for us to appreciate what can be done to protect the precious ozone layer from destruction by human beings.

When substances burn (or when metals tarnish), they combine with oxygen in the air. Yet, even though fire must have been one of humankind's earliest discoveries, and despite the fact that oxygen is the essential component of air that gives us life, it is a mere two hundred years since oxygen was discovered. That discovery was an important step forward because it explained how things burn and was a turning point in the study of chemistry. It also helped to extend chemistry into the realms of biology, marking an important stage in the development of modern biochemistry, the study of the chemistry of life. Indeed, it has been said that the history of oxygen is the history of life.

Remarkably, the significance of the discovery of oxygen was incompletely appreciated by some of the scientists involved and it took the genius of a brilliant French chemist, Antoine Laurent Lavoisier (1743–1794), to explain properly the available experimental observations. Lavoisier's work on oxygen and its extension into other areas of chemistry had such important implications that he is now known as the Father of Modern Chemistry.

The importance of oxygen for life on Earth is taken for granted by most people today. Oxygen is also widely used industrially, for example in the manufacture of steel, and medically, in oxygen tents, oxygen inhalators and incubators for newborn and premature babies. As a rocket fuel, liquid oxygen is helping humankind to explore worlds beyond Earth.

The discovery of oxygen represents an excellent example of how existing ideas in science must be relinquished for new ones when experimental findings no longer agree with old theories. It also demonstrates how important it was for chemists, towards the end of the eighteenth century, to develop ways of measuring the amounts of substances before and after chemical reactions. Precise measurement of the weights of chemicals and how they change during chemical processes was one of Lavoisier's major contributions to science. Perhaps more significant, however, is that the discovery of oxygen and its implications for modern chemistry, medicine and technology originated in the curiosity of scientists and in their quest for an understanding of the natural world.

Phlogiston and the discovery of oxygen

Fire held such an important place in human existence that it was considered by the ancient Greek philosophers, particularly Aristotle (384–322 BC), to be one of the four fundamental elements, along with earth, air and water, which made up the world. A fifth element, ether, was considered to be the substance of the stars. Flames clearly 'escape' from a burning object and the idea arose that burning involves the release of a 'fire-substance' from the object. The German physician, Georg Ernst Stahl (1660–1734), extended this idea and expressed it in his **phlogiston theory** in 1702. Phlogiston (from the Greek, 'to set on fire') was, according to Stahl, present in all burnable objects and it was released into the air when they burned. Substances that burn well were thought to contain lots of phlogiston, whereas those that burn poorly were said to possess small amounts of phlogiston.

The phlogiston theory was useful because it explained many observations and involved a unifying principle – phlogiston – that was involved in the burning of any material. It agreed with the fact that many substances lose weight when they burn. Wood, for example, gets converted to the lighter charcoal or ash. However, it was also known that some metals gained weight after prolonged heating in air. If they release phlogiston, then surely phlogiston must have negative weight? Most scientists at the time had no trouble with this apparent contradiction: phlogiston was regarded by many to be a chemical principle comparable to light or gravity, rather than a substance that had weight.

Phlogiston theory also did not really explain why air was needed for burning, although many scientists explained this by proposing that air is needed to absorb and disperse the phlogiston released during burning.

The phlogiston theory was, like many scientific theories, useful in its day, but it had to be overthrown by new theories that were better able to explain the available experimental evidence. This is very much a part of the scientific process and many theories that have been eventually discarded were essential for scientific and human progress. Theories lead to predictions, which can be tested experimentally, and new experimental results lead to formulation of new theories. Thus science progresses: it is not that theories are useless, but rather that they need constantly to be refined and improved so that they provide an ever clearer picture of the world.

Oxygen was discovered independently by two chemists: Joseph Priestley (1733–1804) in Britain and Carl Wilhelm Scheele (1742–1786) in Sweden. Scheele made the discovery in 1772 but delayed publication of his work until 1777, whereas Priestley made the discovery in 1774 and published his findings in 1775. Priestley was therefore initially given sole credit for the discovery of oxygen.

Priestley was born in Yorkshire, England. His interest in chemistry grew particularly as a result of his experience of teaching science to schoolchildren. In 1758 he opened a day school in Cheshire, England, and made a great success of teaching science, providing the students there with the most up-to-date scientific equipment. He attended lectures and demonstrations in chemistry between 1763 and 1768, and his enthusiasm for science was greatly stimulated when he met Benjamin Franklin in London.

In 1767 Priestley returned to Yorkshire as a church minister. He lived next door to a brewery and there he collected the gas or 'air' released by the fermenting beers. He found that this gas (now known as carbon dioxide) could be dissolved in water to produce a fizzy drink that tasted pleasant. He had, of course, discovered soda water.

Priestley discovered ten other gases in addition to the three (carbon dioxide, hydrogen and air) that were already known at the time. One of these gases, nitrous oxide (laughing gas), later became one of the first anaesthetics to be used in medical surgery. Two years after discovering nitrous oxide, he isolated oxygen.

Priestley's discovery of oxygen came from his observations of a gas given off when, in 1774, he heated mercuric oxide (then known as mercury calx) in a closed container. He noticed that this colourless gas made red-hot wood spark and caused a candle to burn much more brightly than it did in ordinary air. Priestley had already shown that mice eventually die when they are made to breathe ordinary air in a closed container, unless they are supplied with fresh air. He then found that mice survived for much longer periods of time in the new gas obtained by heating mercury calx than they did in ordinary air. He inhaled the gas and received a pleasant sensation. 'My breast felt particularly light and easy for some time afterwards . . . Hitherto only two mice and myself have had the privilege of breathing it', he wrote.

It is easy for us to see a simple interpretation of this discovery in the light of present-day knowledge and in the absence of phlogiston theory. The mercury calx gives off a gas (oxygen) when it is heated and this gas supports burning and animal respiration.

However, Priestley was so strongly influenced by phlogiston that he interpreted his results in a very different way. When mercury calx is heated, he said, it takes up phlogiston from the air. The air becomes dephlogisticated: it now has room for even more phlogiston and therefore supports burning (release of phlogiston from a substance) and animal respiration (production of phlogiston and its release in the animal's exhaled breath) more readily. Priestley's interpretation of the heating of mercury calx can be summarised thus:

Mercury calx + common air → dephlogisticated air + mercury

With phlogiston on his mind, Priestley had no need to invoke the existence of a new gas (oxygen): he believed that his results could be explained in terms of existing ideas: 'common air' around us contains some phlogiston but still has enough room for more, allowing objects to burn and animals to breathe (release their phlogiston). When animals breathe (or a candle burns) in common air inside a closed container, they eventually saturate it with phlogiston. This 'phlogisticated air' can absorb very little extra phlogiston and therefore it does not support burning or breathing. 'Dephlogisticated air', however, from which phlogiston has been removed by heating mercury calx in it, has plenty of room for more phlogiston and therefore supports burning and respiration very well.

Perhaps the real problem with phlogiston was not that it failed to explain the experimental observations, but rather that it was not the simplest explanation. Scientists usually use **'Occam's razor'** to decide between theories that fit all the experimental results. They use the simplest theory that fits all of the evidence. Occam's razor is named after William of Occam (1285–1349), an English philosopher, who was a strong advocate of the principle. Phlogiston made matters complicated, and Antoine Lavoisier came up with a much simpler theory that explained Priestley's observations and the existing knowledge about burning and respiration, that still holds today.

Lavoisier and the death of phlogiston

In October 1774, three months after he had discovered oxygen (dephlogisticated air), Priestley met Lavoisier in Paris whilst visiting the French Academy of Sciences. He told Lavoisier about his work on mercury calx and about the new type of air it produced when it was heated.

Lavoisier already had a deep interest in the process of burning and he also knew that mercury calx lost weight and was converted to mercury metal when it was heated. Indeed, in 1772, Lavoisier had begun to work on the burning of various substances. He had already shown that phosphorus and sulphur, amongst other substances, not only required air for burning but also gained weight in the process. He developed the theory in 1772 that these substances were reacting with, and extracting something from, the air. This theory also explained why metals, including mercury and lead, increased in weight when they were burned in air.

Lavoisier's meeting with Priestley was very significant to Lavoisier because it further stimulated his interest in burning and provided him with some extra information – the production of dephlogisticated air from mercury calx – that helped him to explain even more clearly how substances burn. He immediately set to work on mercury calx. He showed that the production of mercury calx from mercury involved removal of the respirable or burnable part of air, whereas when mercury calx was heated, it produced respirable or burnable air, as Priestley had discovered. Further experiments with tin calx, which was easier to handle, showed that there was no change in overall weight of the air-plus-metal during the burning process. However, the metal gained weight whilst the air lost weight. Indeed, the loss in weight of the air was equal to the gain in weight by the metal: the metal therefore appeared to be removing something from the air.

Lavoisier's explanation of the results he obtained avoided phlogiston. When metallic mercury burns, he said, it combines with part of the air to produce mercury calx. When mercury calx is heated it gives off this same component of the air and is converted back to metallic mercury. He named the constituent of the air '***principe oxygine***', which we now call oxygen:

$$\text{Mercury metal} + \text{Oxygen} \rightarrow \text{Mercury calx}$$
$$\text{Mercury calx} \rightarrow \text{Mercury metal} + \text{Oxygen}$$

The word, 'oxygen' derives from the Greek, meaning 'to produce an acid'. Lavoisier erroneously believed that oxygen was a component of all acids.

Lavoisier deduced that air consists of two major constituents, oxygen and another component we now call nitrogen, which accounts for four-fifths of the volume of the air.

Instead of giving off phlogiston into the air, Lavoisier said, burning substances take up and react with something (oxygen) from the air. There was no requirement for anything with a negative weight – there was no need for phlogiston! However, although Lavoisier's oxygen theory was simple and explained the experimental facts, many scientists were reluctant to accept it at first, and Priestley adhered to phlogiston to the end. Priestley discovered oxygen, but Lavoisier explained his results correctly.

Lavoisier's clear thinking was similarly important when he interpreted the experimental results of others regarding the nature of water. He was the first scientist to infer that water is a combination of hydrogen and oxygen, another monumental step forward for chemistry.

Lavoisier emphasised the importance of measuring weights and volumes before and after chemical reactions, and his studies led him to a fundamental law of nature, the **Law of Conservation of Mass (or Matter)**. 'In every operation an equal quantity of matter exists before and after the operation', he said. This law was of tremendous importance to subsequent studies of chemistry and physics, although it was later refined into the Law of Conservation of Matter and Energy as a consequence of Einstein's Theory of Relativity (Chapter 6). Thus, matter and energy cannot be created or destroyed, but they can be converted one into the other.

Lavoisier published his chemical studies and ideas in a book entitled, *Elementary Treatise in Chemistry*, in 1789. This volume is considered to be a monumental work that began a new era of chemistry. It explained the strengths of oxygen theory and the weaknesses of phlogiston theory. It also did away with many of the old alchemical terms that caused confusion amongst chemists. Lavoisier introduced a new standardised chemical nomenclature that is the foundation of today's chemistry.

Joseph Priestley lived through the French and American Revolutions and was a politically controversial figure in Britain. He held religious views that caused him much trouble. He had been brought up in Yorkshire as a Calvinist and later became a dissenting minister: like Michael Faraday, Priestley held religious opinions that did not conform to the Church of England. Although he was a pious man, Priestley questioned many of the doctrines of Christianity and his reputation as an antagonist of established Christian beliefs was enhanced by his publication of a book entitled, *History of the Corruption of Christianity*, in 1782.

Priestley was seen by those who stood for 'the Church of England, the Country and the King' as a troublemaker. In 1792 a mob of his opponents went wild and destroyed his home, his library, his laboratory and his church. He moved to London but suffered antagonism even there. Eventually, in 1794, he emigrated to the USA, where he was highly respected as a scientist and intellectual. He had tea with George Washington (1732–1799) and became a friend of two other presidents, John Adams (1735–1826) and Thomas Jefferson (1743–1826). He turned down offers of several academic and religious posts, including a professorship in chemistry at the University of Pennsylvania, and spent the last ten years of his life in retirement.

In the same year that Priestley fled from British persecution, Lavoisier met with an even worse fate. He had pursued two careers, one as a scientist, the other as a tax collector under Louis XVI of France. Five years after the French Revolution of 1789, Lavoisier was arrested by pro-revolutionaries along with other tax collectors. In 1794 he was condemned to death in a mass trial and executed on the same day. Of his sentence, someone said, 'The State has no need for intellectuals.' His remains were thrown into a mass grave. The world owes him so much: he revolutionised scientific thought and gave us oxygen and its multitude of implications. The French mathematician, Joseph-Louis Lagrange (1736–1813), said, 'It needed but a moment to sever that head, and perhaps a century will not be long enough to produce another like it.'

11
Of beer, vinegar, milk, silk and germs

Although Antoine Lavoisier, the Father of Modern Chemistry, tragically suffered the lethal blade of the guillotine and did not live to see the dawn of the nineteenth century, the work that he and other chemists of the eighteenth century carried out, especially on oxygen, encouraged the study of the chemistry of life. By the early nineteenth century many scientists were abandoning the old idea that a 'vital' principle existed that was peculiar to living systems. Life, some scientists now believed, was after all based on the same chemical and physical principles as non-living materials and was therefore amenable to physical and chemical analysis.

The belief in an obscure 'vital' principle was one of the factors that kept biological advances well behind those of chemistry and physics, which had progressed tremendously by the beginning of the nineteenth century. By the end of the nineteenth century, however, biology also was on a firm scientific footing as a result of several gigantic strides in biology. Darwin (1809–1882) postulated the theory of natural selection and evolution which revolutionised our understanding of the origin of animal and plant species. The cell theory, which asserted that cells are the basic units of all living things, changed the way scientists saw the workings of plants and animals. Mendel's laws of inheritance, which explained many of the facets of how genetic traits were passed down from parents to offspring, were discovered. Spontaneous generation, the idea that microbes and other creatures, such as maggots, were created out of decaying animal and vegetable matter was refuted, and the germ theory of disease was finally accepted. Of these great advances, the refutation of spontaneous generation and the germ theory of disease were especially important for subsequent medical advances in understanding, preventing and treating bacterial and viral diseases, and one of the most brilliant scientists of all time played a major role in bringing them about. His name was Louis Pasteur.

The achievements of the French chemist, Louis Pasteur (1822–1895), were so numerous and important for human progress that it is truly impossible to give them enough credit in written words. Every day in every way our lives are influenced by this great man's discoveries. Modern surgery can be carried out safely, without risk of infection, because of Pasteur's work. Antibiotics can be used to cure us of bacterial infections as a result of Pasteur's research. Vaccination against viral diseases owes much to Pasteur. Every sewage and water supply system, indeed every form of sanitation one can imagine, is influenced in some way by Louis Pasteur. Even the wine, beer and milk that we drink and the processed food that we eat owe much of their quality to Pasteur. Louis Pasteur stands magnificently and immortally amongst the handful of human beings who have saved humanity from the misery of infectious diseases. His contributions to medicine are amongst the most profound ever made, yet he was never trained in medicine.

Pasteur founded modern microbiology and at the same time

abolished ancient and persistent ideas about diseases that seem to us today to be ridiculous. These old ideas hindered progress of our knowledge of bacterial diseases, and only after Pasteur had set the record straight were significant advances made in the treatment and prevention of infectious diseases that had plagued humankind for centuries. The fruits of Pasteur's work, which are so important to every human being on the face of the Earth, developed out of his deep curiosity and his quest to understand Nature. Pasteur's work demonstrates to all of us that human devastation, disease and death can be alleviated as a result of a scientist's pursuit of pure knowledge. The life of Louis Pasteur stands supremely as a testimony of the fact that the best and most far-reaching applications of science come from passionate studies of seemingly esoteric subjects.

It was obvious to Pasteur and his contemporaries that human diseases existed as separate entities with particular characteristics (symptoms), and that many of these diseases (the 'contagions') were transmissible from one person to another. Everyone knew about the horrifying symptoms associated with epidemics of plague, typhoid, tuberculosis, cholera, smallpox and other pestilences that so frequently devastated humankind. Two of Pasteur's five children died of typhoid fever, and this was not an uncommon experience in the nineteenth century: fatal infectious diseases were a tragic fact of life. Even in today's world, infectious diseases about which we know a great deal are major killers in underdeveloped parts of the world, and the knowledge contributed by Pasteur and other great scientists has yet to be applied in these unfortunate regions.

Research on the process of **fermentation** was closely connected with the germ theory of disease. Pasteur's studies of the microbes that ferment sugar to alcohol, and those that sour milk and wine, led him to the idea that human diseases might bear some similarities to fermentation. Micro-organisms had been well documented, even classified into groups, for almost two centuries before they were established as the cause of contagious diseases. Another concept that was popular before Pasteur's work was that of **spontaneous generation**, the theory that living organisms could be created, without the need for any parent, from decaying animal or plant material. Pasteur demonstrated that this was not possible and showed how the multiplication of micro-organisms might be controlled.

Let us examine the important areas of research that led along the winding road towards establishment of the germ theory of disease.

Microscopic life

Until the microscope was invented in the Netherlands in the late sixteenth century, microscopic life forms, invisible to the naked eye, existed only in the realms of speculation and imagination. The Dutchman, Anton van Leeuwenhoek (1632–1723), had enormous influence on future scientists who studied life under the microscope. Van Leeuwenhoek was a linen draper who later became a local government official in Delft. His main hobby involved making lenses and examining whatever material he could obtain under a magnifying glass made from one of his lenses. His lenses were considered to be perhaps the best in Europe. Beginning in 1673, van Leeuwenhoek studied thousands of substances over a period of more than ten years under his single-lens 'microscope'. He discovered a whole new world of micro-organisms, comprising many different microbial species.

Van Leeuwenhoek regularly communicated his observations to the Royal Society of London, which then published them. He recorded his findings in meticulous detail. For example, in one communication he described the appearance of material (plaque) taken from between his own teeth, 'Though my teeth are kept usually very clean, nevertheless when I view them in a magnifying glass, I find growing between them a little white matter . . . In this substance I judged there might probably be living creatures.' In his own plaque he discovered 'very many small living animals which moved themselves very extravagantly . . . they darted through the water.' Nobody who subsequently re-examined these organisms under the microscope doubted that they were alive, and van Leeuwenhoek's observations were confirmed and extended by many scientists. This whole new realm of microbes was made known to the world largely because of van Leeuwenhoek's hobby.

By the middle of the nineteenth century better microscopes had been perfected and the existence of living micro-organisms was fully established. Yet few scientists believed that microbes were responsible for disease. Perhaps the idea of an invisible organism challenging a sizeable human being was a little too far-fetched for most people to believe.

Once scientists accepted the existence of life forms that were too small to be seen with the unaided eye, they began to wonder whether these microbes might be involved in certain well known, but little understood, processes, such as fermentation. The production of alcohol from sugar, wine from grapes and beer from fermented barley (malt) had been known for centuries to require yeast. Could yeast, some scientists asked, be a living micro-organism?

Fermentation

Agents such as yeast that appeared to be necessary for fermentation were called 'ferments', and one of the big issues in the middle of the nineteenth century was whether or not ferments were living. In the early nineteenth century a number of eminent chemists, including Faraday's mentor, Humphry Davy, had shown that some chemical reactions occurred with greatly increased speed if other agents were present during the reaction, but these added substances were not depleted or altered during the reaction. For example, starch could be converted to sugar by adding sulphuric acid to it and heating the mixture, whereas if the sulphuric acid was omitted the starch could not be converted to sugar. Yet the sulphuric acid was completely unchanged during this process: there was as much sulphuric acid present after the starch had been converted to sugar as there was before the conversion occurred. Likewise, powdered platinum caused hydrogen peroxide to decompose into water and oxygen, even though the platinum was unaltered during the process. The sulphuric acid and platinum used in these studies, and other reagents that increased the rates of chemical changes without themselves being altered, became known as **catalysts**.

Yeast did not appear to be used up during the process of alcoholic fermentation, but it did seem to be necessary for fermentation to occur efficiently. Leading chemists of the time suggested that yeast was simply a catalyst for the conversion of sugar to alcohol. In addition to yeast (the alcoholic ferment) several other ferments were known, including the one that converted wine to vinegar (the acetic acid ferment). These ferments, according to the great Jons Jacob Berzelius (1779–1848), were nothing more than inorganic, non-living chemical catalysts. Another eminent chemist, Justus von Liebig (1803–1873), accepted that yeast might be a living microbe, but he still believed that the ferment responsible for alcohol production was a non-living catalyst that was released from dead or dying yeast organisms.

It was partly due to the opinions of the great chemists of the time that there was a general reluctance to accept the idea that ferments were micro-organisms. Textbooks of the day stated that ferments were non-living catalysts. In addition, the idea that life was involved in fermentation appeared to some scientists to be a step backwards. After all, chemists were already making great strides in understanding biology as a result of their abandonment of 'vital' principles, so why invoke a 'vital' component for fermentation when it could be explained chemically?

When it became clear that ferments were present in stomach juices and in extracts of barley and other living organisms, the chemists simply said that living things contain many of these chemical catalysts. That did not alter their belief that yeast and other ferments were not living organisms. It was one thing to say that ferments were living and another thing to say that living things contained ferments.

When it came to alcoholic fementation, the main problem was whether or not the microscopic oval bodies (which we now know to be yeast cells) present in yeast were living organisms or not. Despite excellent studies by several scientists showing that yeast globules multiplied, and that their degree of multiplication was directly related to the level of fermentation observed, the leading chemists of the day dominated the situation and few scientists were prepared to challenge their authority.

When Pasteur began to study fermentation, a variety of ferments were known, and other processes, such as the decay of meat and eggs, were thought to be due to ferments that produced foul-smelling end-products. It was Pasteur who dared to challenge, with intense conviction, the ideas of the great chemists, and it was he who finally convinced the world, through carefully designed and meticulously performed experiments, that yeast and other ferments were indeed microbes.

Pasteur, ferments and microbes

Louis Pasteur was born in 1822 in Dole, France and trained as a chemist. His early work involved studies of the phenomenon of **molecular asymmetry**, which he discovered. Some molecules exist as two mirror image forms, much the same as left-handed and right-handed gloves, and the two forms often are chemically indistinguishable. However, they do differ in their ability to rotate polarised light: one form rotates it anticlockwise, the other rotates it clockwise. Pasteur discovered that crystals of tartaric acid exist as mirror image forms and that the two forms rotate polarised light in opposite directions. He also demonstrated that only one of the forms of tartaric acid could be metabolised by living organisms: the mirror image was not used at all by living organisms. The discovery that living organisms have an ability to discriminate between mirror image forms of a molecule had a huge impact on our subsequent understanding of biochemical systems.

Pasteur was aware of the processes of fermentation several years before he began to carry out his own experiments on them. It is likely that his intellectual interest in ferments began when he learned that some ferments produce amyl alcohol, which, like tartaric acid, can exist as mirror image molecular forms. Of particular interest to Pasteur was the fact that the amyl alcohol produced during fermentation existed in only one of the two possible forms. It may well be that Pasteur realised the significance of this result from the start and that it established his belief that fermentation was due to living organisms, since he knew that living cells, but not ordinary chemical reactions, were able to discriminate between mirror image forms of the same chemical substance. Throughout the duration of his research on ferments, Pasteur believed that they were due to living organisms, despite the fact that eminent chemists of the time rejected this idea.

Pasteur's experiments on fermentation began in earnest in 1855, a year after he became a professor of chemistry at Lille, France. The father of one of his students owned a distillery in which sugar-beet was fermented to alcohol, using yeast, and was having problems with the process. He went to see Pasteur for advice, and Pasteur decided to examine the ferments in order to see if he could help. When Pasteur saw the yeast globules under the microscope and noticed how their shape changed when the fermentation process became impaired, his interest in ferments was greatly stimulated. He began to research fermentation in depth and must surely have embarked upon these studies with the notion that yeast and other ferments were living microbes.

Within a few years Pasteur had carried out detailed and careful studies on various ferments, including those that produced alcohol, lactic acid, tartaric acid and butyric acid. He discovered the lactic ferment, which caused milk to turn sour: it consisted of microscopic globules not dissimilar in shape from yeast globules, though much smaller. Nobody had seen any obvious microscopic ferment in souring milk before; Pasteur was careful and observant enough to notice the tiny globules of the lactic ferment. He isolated these globules and showed that they readily soured milk, and he discovered that they behaved similarly to yeast in other ways.

When he grew ferments in nutrient broths, Pasteur noticed that they multiplied rapidly and that the offspring resembled their parents, and he demonstrated that their growth paralleled the extent of fermentation. He also showed that onion juice killed some ferments but not others: this might well have been, in the Great Master's eyes, a hint of the antiseptic methods that were to be a consequence of his studies and to revolutionise surgical operations. But these studies alone did not establish the notion that ferments were alive: von Liebig, the chemist, had already proposed that ferments were non-living substances released from dying microbes.

The crucial experiments that Pasteur performed subsequently did convince most scientists that ferments were microbes. He showed, first, that fermentation was not quite as simple as was previously thought. When he examined the products of alcoholic fermentation, for example, he discovered various other chemicals, including glycerine and succinic acid, in addition to alcohol. If Berzelius and his followers were correct that alcoholic fermentation was due to a chemical catalyst that enhanced a simple chemical reaction (the conversion of sugar to alcohol), then why were these multiple chemicals, which could not have been produced from sugar in a simple way, also produced? Pasteur linked the complexity of the products of fermentation with the idea that a complex living organism (yeast) was responsible for the process. He also discovered that many of the additional chemicals produced during fermentation showed molecular asymmetry and that only one of the two possible mirror image forms was present: this was, he believed, evidence for the involvement of life.

More importantly, Pasteur achieved a great feat when he managed to grow yeast in a culture medium made from only sugar, ammonia and simple inorganic salts. Under these conditions, a small number of yeast globules could multiply and use up the components of the medium in the process. Their growth rate paralleled the production of alcohol. Fermentation was definitely not the result of yeast decomposition or death, for here the yeast globules were multiplying beautifully and alcohol was being produced in quantity! The yeast was taking up simple chemicals from the medium and making proteins and other components of yeast from these medium constituents, whereas a non-living catalyst remains unaltered during the chemical process that it catalyses. All of these experimental results agreed with the idea that yeast globules were alive and raised serious arguments against the inorganic theories of ferments.

Pasteur also discovered the butyric acid ferment. This ferment consists of a microbe that moves: it was unquestionably alive because it wriggled and swam rapidly through its culture medium. Many of its features were similar to those he had found to characterise the other ferments that he studied. Butyric acid fermentation occurred more readily in the absence of oxygen than in its presence, and Pasteur is credited as being the first scientist to show that some organisms multiply and thrive in the absence of oxygen.

Pasteur's detailed and elegant studies of ferments established the precise nutrient requirements for growth of the microbes that carry out fermentation processes. Each type of fermenting microbe has its own specific preferences of acidity and alkalinity, growth temperature and nutrient conditions, and each ferment produces its own characteristic products. Through his thorough studies of these ferments, he contributed very important information for the beer, vinegar, wine and alcohol industries, which to this day remain indebted to him for pointing the way towards development of the optimal conditions required for the various fermentation processes. His work showed clearly that the microbes of fermentation could be readily controlled: microbes had become domesticated.

Pasteur's name lives on in our everyday lives in the process of 'pasteurisation', by which milk, wine, and other drinks and foodstuffs are heated to prevent growth of micro-organisms in them. Pasteur applied this method to the wine industry in the 1850s and made a significant contribution towards reducing bacterial infections in wine and improving its taste and quality.

It became clear that fermentation processes were occasionally subject to 'diseases'. Pasteur recognised that these diseases may be due to the presence of other micro-organisms contaminating the fermenting microbes, or to incorrect growth conditions that caused the fermenting microbes to produce the wrong end-products. There is no doubt that during these studies Pasteur was beginning to see a parallel between microbes of ferments and human diseases. If particular types of micro-organisms could carry out specific fermentations or putrefaction, and if other microbes could cause disease of fermentation processes, was it not possible also that some animal diseases were caused by infections with specific types of bacteria? Before he tackled the problem of infectious diseases, Pasteur next turned his experiments to another controversial idea, namely that of spontaneous generation.

Spontaneous generation

The idea of spontaneous generation is a very old one. Even the ancient Greek philosopher, Aristotle, believed that many plants and small animals were formed from decaying soil or putrefying flesh and excrement. The general idea arose that larger animals and human beings once did spontaneously arise from earth, but are now able to breed and reproduce without the need for spontaneous generation. The appearance of maggots, flies, worms, mice, rats and the like amongst decaying animal or vegetable matter is something with which everyone is familiar, but today the idea that worms or mice arise from this material by spontaneous creation is completely bizarre. However, without twentieth-century knowledge, the ancient Greeks and many later European scientists were quite prepared to believe in spontaneous generation.

As late as the sixteenth century, the eminent chemist, Jan Baptista van Helmont (1579–1644), made one of the few scientific mistakes of his life when he claimed to have demonstrated that mice could be generated spontaneously from dirty linen that was placed in an open container with a piece of wheat or cheese. Another great scientist, Francesco Redi (1626–1697), did some better controlled experiments in 1684 and demonstrated that maggots did not appear spontaneously on decaying meat if the meat was covered with gauze, although he observed that flies laid their eggs on top of the gauze and that these eggs developed into maggots. In other words, maggots came from flies, and not from rotting meat.

In the late seventeenth century, just when the evidence against spontaneous generation was accumulating, it raised its mischievous head once more, as a result of van Leeuwenhoek's observations of microscopic living beings. It was accepted that mice and maggots might not, after all, be created daily from the dust of the Earth, but many scientists thought that spontaneous generation was the only plausible means by which the microscopic organisms could appear. How could just one of these microbes produce more than a million offspring in a single day? It was much more reasonable at the time to think that each microbe was spontaneously created, although we know today that bacteria do indeed multiply extremely rapidly. The discovery of micro-organisms started a debate between scientists who believed that they were generated spontaneously and those who believed they had microbial parents in the same way as larger creatures.

Some observations seemed to support the concept of spontaneous generation of microbes. For example, it was known that yeast appeared in huge numbers during the fermentation of grape juice to wine, even if no yeast had originally been added to the grapes. We know now that grapes naturally contain yeast on their skin and that this multiplies when grapes are fermented: it does not appear spontaneously. Other observations, such as the fact that prolonged heating of meat or other putrescible substances prevented them from decaying if they were kept inside a closed vessel free from the air, seemed contrary to the idea of spontaneous generation. When air was admitted, they then decayed. Proponents of spontaneous generation argued that air was necessary for it to occur, whereas opponents argued that air carried with it the microbes that caused putrefaction.

Experiments designed to examine spontaneous generation sometimes agreed with the theory, whereas at other times seemingly similar experiments disagreed with the theory. One of the problems was a lack of reproducibility of experimental results, undoubtedly due to the widespread occurrence of bacteria in the environment and the difficulty of keeping apparatus sterile. How could a culture flask be used to examine whether meat decays under certain conditions if the flask itself, or its lid, is teeming with its own bacteria? Another problem was the lack of care in designing the experiments. If air, for example, was a requirement for spontaneous generation, then the lid of any culture vessel needed to be sufficiently porous to allow air to enter but sufficiently impermeable to prevent entry of microbes. If spontaneous generation did occur, a putrescible substance (from which all microbes had been removed) should have produced microbes if it was placed in the sterile vessel, provided that air, but not fresh microbes, was allowed to enter the flask. If putrefaction failed to occur under these circumstances, nobody could have argued that there was a problem with the entry of air into the vessel, and so the notion of spontaneous generation would have had to be discarded.

Before Pasteur, an Italian scientist, Lazzaro Spallanzani (1729–1799), came nearer than anyone to disproving spontaneous generation. In 1765 he found that by heating vegetables before they were allowed to decay, the number of microbes that subsequently appeared in them was reduced compared with non-heated vegetables. More importantly, the vegetables had to be heated for almost an hour in order to be sure that microbes did not subsequently grow in them. This experiment showed that it was not as easy to eliminate existing microbes from substances by heating as was previously thought. Spallanzani also placed vegetable matter in glass vessels and examined the effect of various ways of sealing the vessels on the ability of the vegetable matter to grow microbes. He discovered that more microbes grew on the vegetation as the porosity of the lids increased. Open vessels produced the greatest number of microbes, whereas there were fewer microbes as the lid progressed from cotton to wood to glass. Spallanzani interpreted this result to mean that air contains microbes and that these microbes gain access to the glass vessel more easily as the porosity of the lid increases. If the vessels are properly sealed, the vegetables fail to grow microbes, indicating that the micro-organisms are not produced spontaneously, but when they do appear it is because they are brought into the vessels from the air. The proponents of spontaneous generation still argued that it was not the microbes themselves that were present in the air, but some other component of air that was needed for spontaneous generation to occur: as the porosity of the lid fell, less of this component of air was able to enter the vessels.

The solution to the spontaneous generation problem was proving to be a difficult one. Every time the opponents of the theory came up with an argument against it, the proponents had a counteractive argument in its favour. It would take a great mind and a superb experimentalist to solve the problem once and for all. Louis Pasteur was the scientist who provided the crucial evidence that sent the spontaneous generation theory into oblivion from its two thousand-year existence.

Pasteur and the end of spontaneous generation

Pasteur's verification that ferments were living microbes ended one scientific argument and he next turned to the great controversy of spontaneous generation. In 1860 he wrote; 'I am hoping to make a decisive step very soon by solving, without the least confusion, the celebrated question of spontaneous generation.' It appeared that he knew exactly what experiments were needed to destroy once and for all the idea of spontaneous generation, even though many scientists before him had tried and failed to provide the decisive blow. Pasteur's superiors, who by now had tremendous regard for his scientific and intellectual ability, were concerned when he told them that he was about to take up the challenge of spontaneous generation. They warned him that he would become engrossed in difficult experiments that would not prove conclusive enough to convince opponents of the theory. Pasteur carried on regardless: perhaps he knew that he had to tackle the problem as a necessary step towards proving the germ theory of disease.

One of the major issues to address was whether the microbes that appeared in culture broths were brought there by the air, or whether the air was merely required for their spontaneous creation in the broths. Pasteur filtered air, collected the debris and examined it under the microscope. He found that air contained small globular particles very similar to those he had seen growing profusely in culture broths, and he showed that these particles were able to grow in the broths. Germs were certainly present in the air, but this still did not prove that spontaneous generation was not responsible for growth of microbes in nutrient broths.

The real evidence came from a series of astonishingly simple but elegant experiments. Pasteur poured some nutrient broth into a number of glass vessels and he teased out the necks of these flasks into long thin curves of various shapes (Figure 20). These 'swan-necked flasks' were then boiled so that the air in them was forced out. When they cooled down, the flasks of broth were placed to one side with similarly boiled flasks in which the necks were not curved. In all flasks the outside air gained free access to the broth, since none of the necks had been sealed but any microbes might be trapped in the curved necks. The results of this simple experiment were conclusive. No microbes grew in the swan-necked flasks, whereas heavy infections grew in the flasks with straight necks. When Pasteur broke the curved necks on some of the swan-necked flasks he found that bacteria then grew in these same flasks. Indeed, some of Pasteur's original swan-necked flasks containing broth that was boiled more than a hundred years ago are still in existence, and they are free of microbes to this day!

Pasteur's swan-necked flasks answered the criticisms that many proponents of spontaneous generation had raised previously. Here were flasks containing broth that was directly in contact with the air: no lids were used, and no microbes were spontaneously created. His experiments proved that for bacteria to grow in a broth they had to gain access to it from the surroundings. Whilst air could freely enter and leave the swan-necked flasks, the heavier bacteria became trapped in the curved portions of the necks and never reached the broth. Pasteur placed a drop of previously boiled broth in the curved neck of one of the flasks and found that it did grow bacteria: the microbes from the air did indeed become trapped in the curved necks.

Pasteur also showed that bacteria in air vary in number from one place to another. When he exposed previously boiled broth to air at high altitudes, for example in the Swiss glaciers, the broth remained sterile. In the face of opposition to his ideas, he eventually persuaded the French Academy of Sciences to appoint a committee to repeat his experiments so that they could be verified. His confidence in his own data was unfailing, whereas his opponents withdrew their opposition, obviously because of their lack of certainty in their own data. Spontaneous generation was vanquished once and for all. Whereas there once was a time

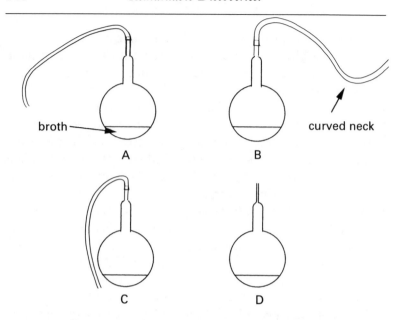

Figure 20.   Flasks used by Pasteur to refute the spontaneous generation theory. Flasks A, B and C each had curved necks and allowed air, but not bacteria, to enter: the broth in these flasks remained sterile. Flask D lacked a curved neck: its broth became infected with bacteria from the air.

when few dared to challenge the idea, now nobody would venture to oppose Pasteur's extermination of it.

The germ theory of disease

Pasteur had progressed from chemistry to fermentation then on to spontaneous generation with breathtaking success. It was an inevitable step for him next to apply his 'germ theory' of fermentation to a 'germ theory' of disease. In 1862 he wrote, 'The study of germs offers so many connections with the diseases of animals and plants, that it constitutes a first step in the . . . investigation of putrid and contagious diseases.' Fourteen years later he began his own experiments with anthrax bacteria and confirmed the germ theory.

Before the germ theory of contagious disease was finally accepted, there had been centuries of mysticism and misunderstanding regarding the causes of disease. Some people had thought that diseases were divine retribution for corruption or sins. Others believed that there were earthly 'forces' and that diseases were caught and spread when these got out of control. Yet others thought that diseases were caused by 'bad airs', or 'miasma': malaria, for example, was thought to be due to miasma that rose from swamps (Chapter 13). Few considered contagious diseases to be due to minute, invisible living organisms and those who did were unable to provide sufficiently conclusive experimental evidence.

In 1836, the Italian scientist, Agostino Bassi (1773–1856), demonstrated that a fungus was the cause of muscardine, a disease of silkworms that was causing problems to the silk industry. Bassi showed that material taken from white patches that appeared on silkworms with muscardine produced the disease when it was injected into healthy silkworms. He noticed that the infectious material contained a fungus when he examined it under a microscope. It soon became clear that certain other diseases, for example the human disease thrush, were also caused by fungi.

Silkworm diseases also provided Louis Pasteur with his first direct contact with infectious diseases at the experimental level. France's silk industry was being devastated by an epidemic in which commercially bred silkworms were succumbing to diseases in large numbers. In 1865, the Ministry of Agriculture set up an investigation of the silkworm epidemic and Pasteur was chosen to lead the study. The disease involved the appearance of small black specks on the skin of the silkworms and nobody knew the cause. Pasteur began his work on the problem believing that germs were not responsible. He thought the disease was merely a physiological problem of the silkworms, even though microscopic globules ('corpuscles') were found in infected silkworms and the chrysalises and moths that developed from them. However, after several years of experiments, Pasteur was convinced that he had identified micro-organisms as the cause of two diseases, flacherie and pébrine, that plagued the silkworm industry. On the basis of his experimental findings, he advised those working in the silk industry on the best means of eliminating the disease and thus made a significant contribution to the silk industry of France and other European countries.

Pasteur's investigations of infectious diseases of silkworms were his stepping stone to experimental confirmation of the germ theory of disease in larger animals and humans. In 1876, and in spite of the fact that he was a sick man, having suffered a stroke that paralysed him down his left side, Pasteur decided to carry out an experimental investigation of anthrax, a terrible disease that was decimating France's sheep population. More than a decade earlier, others had seen rod-shaped micro-organisms in the blood of sheep with anthrax and had shown that their blood produced anthrax when injected into healthy sheep. However, there had been some controversy about these experiments and, as with spontaneous generation and fermentation, the results obtained were not conclusive enough to satisfy opponents of the germ theory.

When Pasteur began his investigations of anthrax, Robert Koch (1843–1910), a German doctor, was already intensively studying the disease. Koch shares the credit with Pasteur for verifying the germ theory of disease and for giving us today our freedom from many devastating bacterial diseases. Koch grew the microscopic rods found in the blood of anthrax-infected sheep in drops of blood or in fluid taken from the eyes of rabbits. He showed that these rods, now known to be anthrax **bacilli** (rod-shaped bacteria), produced anthrax when they were injected into mice, and he found many more bacilli in the infected mice. Rod-shaped bacteria taken from other sources, for example hay, failed to produce anthrax, and Koch correctly interpreted this to mean that anthrax was caused by a specific type of bacillus. Koch elucidated the whole life history of the anthrax bacillus: he saw them form filaments of rods when they divided, then he noticed that they developed into round structures (spores), which were able to survive harsher conditions than the bacilli.

When Koch grew anthrax bacilli in blood or eye fluid, he made the important step forward of isolating a disease-causing microbe and growing it outside an animal. It was now clear not only that germs caused disease, but also that they could be cultivated and studied in large numbers independently of any animals.

Some scientists were still critical of Koch's work. They believed that anthrax was not caused by the bacillus but by another factor that was transferred with it when blood was removed from an infected animal. Pasteur answered this criticism and complemented Koch's results to establish the germ theory of disease once and for all. Pasteur grew the anthrax bacilli in urine and made serial dilutions of his first culture until any non-multiplying contaminating factor present with the bacilli was diluted out to such an extent that none of it could possibly be left. The culture, in which anthrax bacilli had been continuously growing, still produced massive infections in animals. This proved beyond reasonable doubt that the bacilli did, indeed, cause anthrax. Moreover, when Pasteur filtered out the bacilli from his cultures, infectivity was lost.

Once the germ theory of disease was established, scientists began using the meticulous methods pioneered by Koch and Pasteur to examine the possibility that microbes caused diseases other than anthrax. By the end of the nineteenth century, microorganisms were identified that caused tuberculosis, cholera, malaria and many other diseases. Koch received the 1905 Nobel Prize in Physiology and Medicine for his pioneering work on bacterial diseases; Pasteur died in 1895, before the Nobel Prize was established. It cannot be doubted that if Pasteur had lived he would have shared the Prize with Koch. Indeed, his work on each of the areas of molecular asymmetry, ferments and spontaneous generation was individually of a sufficiently high calibre to warrant a Nobel Prize. Yet Pasteur's stupendous contribution to human progress did not end there: his work on vaccines was yet to come (Chapter 12).

Pasteur's research on the germs that cause fermentation and his ideas about the processes of decay and disease were generally ridiculed by the medical profession. Most medical doctors considered his ideas preposterous: who was this mere chemist who claimed to have the answers to medical problems? However, some medics were beginning to take notice of him. One of them was Joseph Lister (1827–1912), a British surgeon. Lister heard about

Pasteur's work on the microbes that cause putrefaction and he believed, like Pasteur, that micro-organisms were responsible for the appalling infections that frequently occurred in surgical wounds. In the mid nineteenth century surgery was considered to be the last resort in treatment of medical problems: deaths caused by infections resulting from surgical intervention were the rule rather than the exception. Lister reasoned that if these infections were due to microbes, there might be chemical treatments that kill the microbes and so reduce problems following surgery.

In 1864, Lister developed antiseptic methods aimed at killing bacteria at the site of surgical wounds. His success with carbolic acid sprays was enormous: he dramatically reduced the incidence of infections and deaths following surgery. Lister's antiseptic surgery led to today's rigorous standards used in operating theatres, where germs are prevented from gaining access to surgical incisions. There is no doubt that Lister was inspired by Pasteur. In a letter he wrote to Pasteur in 1874, he said:

> 'Allow me to take this opportunity to tender you my most cordial thanks for having, by your brilliant researches, demonstrated to me the truth of the germ theory of putrefaction, and thus furnished me with the principle upon which alone the antiseptic system can be carried out. Should you at any time visit Edinburgh, it would, I believe, give you sincere gratification to see at our hospital how largely mankind is being benefited by your labours. I need hardly add that it would afford me the highest gratification to show you how greatly surgery is indebted to you.' (From Dubos, R., 1960.)

Throughout Pasteur's life he encountered opposition to his theories, but he always triumphed over his antagonists. The main reasons for his success are probably his perseverance, conviction of the correctness of his ideas, the brilliance and simplicity of his experiments and the care with which they were performed. In the end, the results of his experiments were self-evident and nobody could argue with them, and they usually turned out to be correct.

Once it was clear that germs caused disease, and many of the microbes responsible had been isolated, some scientists began to consider the possibility that substances might be made that kill microbes but not the animals infected with them. Lister's antiseptic surgery already demonstrated that wound infections could be prevented by killing germs, and by the end of the nineteenth century it was quite clear that many diseases were associated with unhygienic living conditions and that clean living standards go some way to preventing many contagious diseases. One disease, smallpox, had already been conquered in large measure, as a result of the vaccination method developed by Edward Jenner (Chapter 12), although the mechanism by which smallpox was prevented by vaccination remained obscure. The establishment of the germ theory of disease led, initially by chance, to the production of vaccines against many more contagious diseases. Pasteur played a major role in development of these new vaccines.

12

Of milkmaids, chickens and mad dogs

Towards the end of the eighteenth century, Thomas Jefferson, President of the USA, wrote a letter to the British doctor, Edward Jenner (1749–1823), congratulating Jenner on his discovery of the smallpox vaccine. 'You have erased from the calendar of human afflictions one of its greatest. Yours is the comfortable reflection that Mankind can never forget that you have lived; future generations will know by history only that the loathsome smallpox has existed', wrote Jefferson. Two centuries later, in 1977, President Jefferson's prophecy came true when the World Health Organisation (WHO) announced that smallpox had finally been eradicated from the face of the Earth.

The elimination of smallpox was the result of a monumental

effort on the part of the WHO, world governments and health care workers around the globe. It involved keeping track of every new case of smallpox and vaccinating anybody who had come into contact with these new patients, until all contacts of the last known victim had been vaccinated. In 1967 two million people died world-wide as a result of smallpox, yet since 1977 there have been no reported fatalities from the disease (except for the deaths of several scientists who accidentally caught smallpox in their laboratories soon after the WHO announced the eradication of the disease). Routine vaccination against smallpox is no longer carried out: the world will never see smallpox again.

The eradication of smallpox shows just how powerful vaccines can be in the fight against disease. Yet in 1992, it was estimated that two million children die every year of diseases for which vaccines are already available: that is equal to one child dying every thirty seconds. Science has provided these vaccines, but the poor countries of the world have, for political, economic and other reasons, limited access to them. Tuberculosis, tetanus, measles, whooping cough, poliomyelitis and diphtheria are major child-killers in the late twentieth century in underdeveloped countries, yet vaccines are available for all of them. These six diseases constitute the main targets for the WHO's Extended Programme of Immunisation, which aims to bring these vaccines to every child in the world.

Vaccination against infectious diseases goes back thousands of years, but it was Jenner who introduced the safer forms of modern vaccination in the eighteenth century. Another century passed before any more vaccines were developed and the mechanisms of vaccination were beginning to be understood. Louis Pasteur provided the main impetus that led to the flurry of vaccines produced in the twentieth century. Nowadays we have vaccines for the prevention of many diseases in addition to the six diseases mentioned above, including cholera, yellow fever, hepatitis B, influenza, plague, anthrax and rabies. Another disease, acquired immunodeficiency syndrome (AIDS), has appeared on modern man's list of devastating illnesses, and threatens almost every nation of the Earth. AIDS is already a major problem in some parts of Africa, and scientists are searching intensively for a vaccine against the human immunodeficiency virus (HIV) that causes AIDS.

The history of modern vaccines begins with smallpox, which was once one of the most terrible and common diseases that anyone could contract. Future generations will, mercifully, never see those horrific symptoms again.

Lady Montagu, tribal doctors and variolation

The symptoms of smallpox were well known in ancient times and the disease afflicted both the rich and the poor. Even the mummified remains of Rameses V, the Egyptian pharaoh who lived more than three thousand years ago, show evidence of smallpox infection. The disease began with aches and pains, lethargy, disinterest in food and fever. Then came the rash, which consisted initially of small spots, especially on the face and trunk. The spots subsequently became enlarged with pus producing a horrifying sight, and the fever often became more intense. Many people died of smallpox and in those who survived the spots dried up and scarred, often producing permanent disfigurement of their faces. As many as one person in every five died of smallpox in England before the eighteenth century, and almost everybody eventually caught the disease. Blindness was a common complication of smallpox and half of the population of England had pock-marked faces as a result of smallpox infection.

Smallpox virus survived outside the body for months, even years, and was frequently transmitted from one person to another in laundry or household dust. The colonists who arrived in the USA in the early sixteenth century used smallpox-infested blankets and handkerchiefs as weapons of germ warfare against the American Indians, who were especially susceptible to the disease, having been free of smallpox before the colonists arrived there.

Protection against smallpox was well known before Jenner introduced his vaccination procedure in the late eighteenth century. It was common knowledge that once a person had been infected with smallpox he or she was unlikely to be infected again. Somehow, one infection provided immunity to further infection. The ancient Chinese and Indians used to sniff the drying scabs of smallpox victims: this often provided some protection from infection. Another procedure involved transferring some fluid from the spots of an infected person to a small cut made in the arm of an uninfected person, and this also appeared to produce some immunity to smallpox.

The practice of inoculating someone with fluid from smallpox spots was known as '**variolation**': the Latin word for smallpox is *variola*. Variolation involved transfer of small amounts of smallpox virus and therefore produced a low level smallpox infection in inoculated individuals. Survival of this infection would then protect from future infection. However, a person who had been variolated did actually have the disease and therefore could spread it to other people, and there was a risk that too much virus might have been transferred during variolation, with the result that some inoculated people died of smallpox or were permanently scarred by it.

Variolation was unknown in the west before the seventeenth century, although it had been practised in China, India and the Middle East for some time. It was brought to Britain in 1717 by Lady Mary Wortley Montagu (1689–1762), the wife of the British Ambassador to Turkey. She was one of the many beautiful ladies whose face was pock-marked as a result of smallpox, and she noticed the practice of variolation in Turkey. She subsequently carried out variolation in England and eventually it became a common procedure there. Within a few years, smallpox deaths in London amongst those who had been variolated were ten times less frequent than amongst those who had not been variolated.

Variolation was also practised in Africa amongst the tribes there: the tribal doctor would transfer fluid from the smallpox spots of an infected person to a non-infected person, or from the arm of one variolated person to another. The method was brought to the USA in the early seventeenth century as a result of the slave trade. Cotton Mather (1663–1728), a puritan, discovered that one of his African slaves had been variolated in Africa and Mather introduced the practice in colonial Massachusetts.

Improvements in variolation reduced the smallpox death rate

substantially and by the end of the eighteenth century tens of thousands of people had been variolated with only a few deaths from smallpox occurring amongst them. Smallpox inoculation hospitals were common where variolation was carried out and where those inoculated (who were infectious), could be kept in isolation from the rest of society for several months until the mild symptoms had subsided. Groups of friends would often be variolated together so that they could be with each other during the quarantine period.

Despite the enormous success of variolation, it did have its risks and it also had the disadvantage that a variolated patient had to spend time in quarantine, which meant time away from work and family. Edward Jenner ended the practice and introduced a safer procedure for preventing smallpox that continued to be used until the dreadful disease was eliminated from the planet in 1977.

Jenner, cowpox and vaccination

Edward Jenner was born in Gloucestershire, England, in 1749. As a young child he was inoculated against smallpox by the commonly used practice of variolation. When he was thirteen, his family decided to have him trained as a doctor and he was therefore apprenticed to a surgeon. At twenty-one, Jenner went to London to become a pupil of the great British surgeon, John Hunter (1728–1793), who developed many of the early surgical instruments. Jenner then returned to Gloucestershire to practise as a local doctor.

Amongst his patients, Jenner encountered a milkmaid who, he suspected, had smallpox. She replied that she could not possibly have smallpox because she had already been infected with cowpox as a result of milking cowpox-infected cows. It appeared that an old wives' tale was rife in Gloucestershire which said that milkmaids and cowherds were immune to smallpox once they had become infected with cowpox. Jenner investigated this rumour in more detail and found, indeed, that a number of milkmaids who had been infected with cowpox did fail to develop smallpox. Cowpox was a disease of cows in which spots not unlike those of human smallpox appear on the cows' teats; some believed that it was the cow version of human smallpox. It was transmitted from an infected teat to a cut that was present on a person's hand. The fingers of milkmaids would show cowpox spots very similar to those seen in smallpox, but symptoms were mild and the milkmaids would recover completely within a few days.

Jenner's observations made him certain that milkmaids were, indeed, immune to smallpox. In 1796, he decided to test his theory. He took some fluid from the cowpox spots on the fingers of a farmer's daughter, Sarah Nelmes, who had contracted cowpox by milking one of her father's cows after she had pierced her finger with a thorn. He then transferred some of this fluid, using a clean lancet, to two small cuts he had made in the left arm of an eight-year-old boy called James Phipps. The boy had not been variolated previously and he had never had smallpox. After a few days, young Phipps showed signs of mild fever, but he quickly recovered, just as the milkmaids recovered from cowpox. About two months later, Jenner took some smallpox-containing fluid and inoculated it into Phipps, just as he would have done if he was variolating somebody by the standard practice. As Jenner predicted, Phipps failed to develop any symptoms of smallpox. This was the first time that cowpox fluid had ever been transmitted deliberately from one human being to another, and Phipps was the first person to be vaccinated against smallpox without the use of human smallpox itself as vaccine. The word 'vaccine' comes from the Latin, *vacca*, meaning cow. Jenner wrote,

'Phipps was inoculated in the arm from a pustule on the hand of a young woman who was infected by her master's cows. Having never seen the disease but in its casual way before, that is, when communicated from the cow to the hand of the milker, I was astonished at the close resemblance of the pustules in some of their stages to the variolus [smallpox] pustules. But now listen to the most delightful part of my story. The boy has since been inoculated for the small pox which as I ventured to predict produced no effect.' (From Fisher, R. B., 1990.)

In 1797 Jenner submitted his idea and experimental results on the protection against smallpox by cowpox vaccination to the Royal Society of London for publication. The article was rejected because Jenner simply did not have enough data to support his claim. That cowpox could protect against smallpox was generally considered to be ridiculous anyway. The President of the Royal Society told Jenner that he should not risk his reputation by trying to publish data that was so much at odds with the ideas that were accepted at the time. However, Jenner was so certain of his results that he carried out more successful vaccinations using fluid from cowpox pustules and published his work privately in the form of a short pamphlet.

Cowpox was not very common in Gloucestershire, so Jenner decided to vaccinate people with fluid taken from cowpox pustules of someone who had recently already been vaccinated. By repeating this arm-to-arm vaccination, Jenner found that vaccination could be carried on without the need for individuals who had been newly infected with cowpox as a result of milking cows.

There was a tremendous amount of opposition to Jenner's method of vaccination. In the early days of its use there were some problems: proper vaccination was sometimes not achieved and the patient would later succumb to smallpox, and one hospital accidentally mixed smallpox fluid with cowpox fluid, with devastating consequences. Nevertheless, many doctors reproduced Jenner's results and vaccination using cowpox virus became widely used and spread rapidly and far across the world. Jenner's name became famous everywhere. He sent some of his vaccine to the USA, where it was extensively applied: even President Jefferson had his own family vaccinated. Edward Jenner's vaccination procedure stood the test of time.

Smallpox remained the only disease that could be prevented by vaccination until Louis Pasteur developed similar methods for preventing other diseases as a result of his studies of microbes. One of the most remarkable things about the smallpox vaccine is that it was produced at a time when nobody was aware that infectious diseases were caused by microscopic organisms. The cause of smallpox was unknown, but the disease could be prevented by vaccination.

Louis Pasteur and attenuated vaccines

Louis Pasteur and Robert Koch established the germ theory of disease in the 1870s, especially as a result of their investigations of the anthrax bacillus (Chapter 11). Pasteur began to think in depth about some of the well known features of infectious diseases. Why, for example, did exposure to some diseases under certain circumstances produce immunity to further infection? It was well known that people who survived a smallpox infection rarely suffered any further infections with the disease, and this fact had, of course, been the principle behind variolation. Other diseases, such as measles and chicken-pox, showed similar immunity to further infection. Pasteur also started to consider the mechanism of vaccination, which at the time could be carried out only for smallpox. Would it be possible to develop vaccines against other diseases? At the end of the nineteenth century scientists were debating the nature of Jenner's smallpox vaccine. One group of scientists believed that cowpox and smallpox were distinct and independent diseases and that cowpox protected against smallpox for some unknown reason. The other group thought that cowpox was merely an attenuated form of smallpox, and that cowpox was originally derived from smallpox, but changed its constitution when it was passed through a cow making it less harmful to humans.

These considerations of immunity were certainly on Pasteur's mind and he must surely have been contemplating how he might tackle them experimentally. However, he discovered answers to some of his questions serendipitously. These chance findings turned vaccination into a general phenomenon rather than the isolated one for smallpox that it had been for nearly a hundred years.

Pasteur's first major encounter with potential new types of vaccines occurred when he began to study chicken cholera, a disease that was epidemic amongst farmyard hens and cockerels, and was therefore of concern to French farmers. Chicken cholera caused hens and cockerels to stagger, lose consciousness and die. Healthy chickens contracted the disease by eating food contaminated with the excrement of infected chickens. It had already been shown that the disease is caused by a microscopic organism in much the same way that Pasteur and Koch had demonstrated that anthrax is caused by a microbe.

Pasteur was interested in growing the chicken cholera microorganism in his culture broths and verifying that it did, indeed, cause the disease in hens. He obtained the head of an infected cockerel, which contained the bacterium, and he started to investigate different culture broths for their ability to support growth of the microbe. It became clear to Pasteur that the chicken cholera microbe preferred growth conditions that were quite different from those preferred by other microbes such as the anthrax bacillus, and he eventually managed to grow cholera bacteria in a broth made from chicken gristle. Having found an efficient way of cultivating the microbe in laboratory flasks in the absence of any chickens, he was now in a position to study the microbe in depth.

Pasteur found that chicken cholera microbes grown in his laboratory were highly infectious to chickens, and the infected chickens always contained large numbers of the microbe in their blood. The microbe was, indeed, the cause of chicken cholera. When laboratory grown bacteria were fed by adding fresh broth to the culture flasks, the microbes remained highly infectious to injected chickens. During the summer of 1879, from July to October, Pasteur took a holiday. Before he left the laboratory, he made sure that there were some flasks of culture broth containing chicken cholera microbes, so that he could continue using them after the vacation.

When he returned to his laboratory, Pasteur injected some chickens with the old culture broth containing cholera microbes that had been left standing over the holiday. None of the chickens succumbed to cholera, and Pasteur thought that the bacteria must have died because they had been neglected for too long. He therefore decided to start a fresh culture using a new batch of cholera microbes taken from an infected chicken. He injected this fresh culture into *the same* chickens that he had injected with the old culture and which had failed to become infected. He also injected some new chickens (which had not been injected previously), with the fresh cholera microbe culture. To his surprise, Pasteur found that only a proportion of the chickens he injected with fresh cholera micro-organisms developed the disease: the rest remained unaffected. When he examined his results more carefully, he discovered that the chickens that survived the fresh bacterial culture were the ones that had previously been injected with the 'stale' culture, the one that had been left to stand over the summer vacation. The chickens which had not received the stale microbe culture all died of cholera.

Many scientists would have dismissed this experimental observation as being insignificant, but not Louis Pasteur. He knew enough about Jenner's smallpox vaccine to see a possible parallel between Jenner's vaccination and his own results. Perhaps, he thought, the bacteria in the old cultures had become altered (attenuated) in such a way that they no longer produced cholera, but were able to protect against subsequent infection with the normal virulent strain of cholera. In this case, the attenuated microbe would correspond to Jenner's cowpox vaccine, except that Pasteur knew for sure that the attenuated cholera microbe was derived directly from the virulent, disease-producing form, whereas it was not known if cowpox was derived from smallpox.

Further experiments confirmed that attenuated cholera microbes protected chickens from subsequent infection with virulent bacteria. Pasteur found that attenuation was caused by the presence of air: cultures grown in 'open' flasks (with cotton wool lids) were easily attenuated, whereas those grown in sealed flasks were not. By increasing the length of time between additions of fresh broth, he discovered that increasing degrees of attenuation were achieved: older cultures grown in the presence of air were attenuated to the highest degree.

Pasteur had, for the first time since Jenner developed the smallpox vaccine, produced a vaccine against another disease. It was Pasteur who coined the word 'vaccine' for any agent that provides immunity to infection by a micro-organism; originally, the term was applied specifically to the smallpox vaccine. More importantly, Pasteur saw enormous significance in his work on chicken cholera for future vaccine development. Was it not now possible to attenuate an infectious organism in a culture flask and to use this attenuated organism as a vaccine to protect an animal from the real disease? He immediately turned his studies to diseases other than chicken cholera in the hope that he had discovered a general principle, namely that vaccines can be obtained to many diseases by appropriate manipulation of the infectious agents in the laboratory.

Between 1880 and 1884, Pasteur and his co-workers developed vaccines for three more diseases: anthrax, rabies and swine erysipelas. It took almost a hundred years to progress from smallpox to any other disease, but within the space of four years Pasteur showed that vaccines could be obtained against many diseases. Soon after his great success with chicken cholera, he turned his attention to anthrax.

It was known at the time that anthrax could form highly resistant spores that were able to survive harsh conditions. Anthrax spores were found in the ground in areas where anthrax-infected animal corpses were buried and persisted in virulent form for decades after the animals had been buried. During his research on anthrax, Pasteur made frequent visits to local farms. On one such occasion he noticed something unusual about an enclosed area containing eight sheep with which he was experimenting. Pasteur noticed an area of earth inside the enclosure where the soil was a different colour to the rest of the soil. When he injected these sheep with anthrax culture, some of them failed to succumb to the disease, a rather puzzling discovery in view of the fact that the dose of bacilli given was normally highly lethal. The farmer who owned the land told Pasteur that an anthrax-infected sheep had once been buried in the ground in the area where the soil was discoloured. Pasteur examined earthworms taken from the soil under a microscope: the earth inside these worms contained anthrax spores. Was it possible, Pasteur asked, that the live sheep had eaten grass from this isolated area of soil and had become infected with sub-lethal doses of anthrax bacilli that conferred immunity to subsequent infection? It was at least clear that earthworms can bring anthrax spores to the surface from buried corpses. Pasteur made a significant step forward in controlling the spread of anthrax by informing farmers never to bury dead animals infected with anthrax on land used for pasture because dangerous anthrax spores could appear on the ground and be eaten by grazing animals, which could become infected by ingesting lethal doses of the spores.

It was clear to Pasteur that the method he had used to attenuate the chicken cholera microbe would not work with anthrax: old cultures of anthrax bacilli exposed to the air merely formed the hardy spores. What was needed was a method for preventing spore formation. After a good deal of experimenting, Pasteur and his colleagues found that anthrax bacilli in culture broth failed to grow above a temperature of 45 degrees Centigrade and no spores were formed. However, when the temperature was lowered by a very small increment to 42 or 43 degrees Centigrade, the anthrax bacilli continued to grow in culture broth, but they did not form spores. Growth of anthrax bacilli at 43 degrees Centigrade for a few weeks was sufficient to attenuate the bacilli enough such that they produced only a mild illness when injected into animals: the animals did not die. When animals injected with these attenuated anthrax bacilli were subsequently given lethal doses of virulent anthrax bacilli, they failed to develop the disease.

Pasteur had produced a vaccine against anthrax. Despite his confidence in his team's results, however, there was a great deal of scepticism, particularly from the medical profession and veterinary surgeons. The acid test of Pasteur's vaccine was yet to come, in the form of a public demonstration of its efficacy in which crowds of onlookers observed the whole experiment as if it were a circus act. This famous demonstration occurred in 1881 on a French farm called Pouilly le Fort, and it established the importance of vaccination against diseases other than smallpox once and for all in the eyes of scientists, doctors, vets and the public.

This demonstration of Pasteur's vaccine against anthrax was organised by a veterinary surgeon in conjunction with a local agricultural society. Pasteur was invited to vaccinate, in public, a group of farm animals and subsequently to infect the vaccinated animals and a group of unvaccinated animals with lethal doses of anthrax bacilli. The spectators would gather again some weeks later to examine the animals in order to determine whether the vaccine worked. Pasteur accepted the challenge with his usual confidence in his experimental data, although this time his public reputation was at stake. There was a great deal of scepticism from some people, who believed that Pasteur could not possibly have produced an anthrax vaccine; some still did not accept the idea that anthrax was caused by microbes!

The Pouilly le Fort demonstration was widely publicised. Leaflets were even printed and sent out to encourage people to attend. When the day of vaccination finally arrived, Pasteur and his co-workers were met by large crowds of spectators. Twenty-four sheep, five cows, one ox and one goat were injected under their skin with Pasteur's attenuated anthrax vaccine. Another twenty-four sheep, four cows and a goat were left unvaccinated. Pasteur then gave a lecture at the farm in which he predicted the outcome of the experiment and gave a summary of the methods by which he had arrived at the vaccine.

Twelve days later, the vaccinated animals were given another injection of attenuated anthrax vaccine: laboratory experiments had shown that two injections produced the best immunity to subsequent infection.

After the second injection, all the animals were still alive and the next step involved injection of virulent anthrax bacilli into the animals to produce the disease and to assess the efficacy of the vaccine. Pasteur said, 'If the success is complete, this will be one of the best examples of applied science in this century.' Two weeks after the second inoculation, all vaccinated animals and control (non-vaccinated) animals were injected with lethal doses of anthrax culture grown in Pasteur's laboratory. Scepticism was so strong that some people attending the demonstration suspected that Pasteur might cheat at the final injection by giving vaccinated animals lower doses of anthrax than non-vaccinated animals. Somebody made sure that the test-tube containing the anthrax bacilli was well mixed; others ensured that animals were properly injected, and higher doses of the bacteria were injected than Pasteur had originally intended. Pasteur did not object to any change in protocol if it satisfied the spectators that everything was done honestly and properly. In that way, if his predictions were verified nobody could criticise his anthrax vaccine.

Pasteur was admittedly nervous following the injection of lethal anthrax into the animals: his reputation and that of his vaccine was at stake and the outcome of the demonstration would be known within a few days, when infected animals should start to die of anthrax. The day after infection, he received news that some of the vaccinated animals were sick and this obviously made him uneasy, but later the same day further news came that the animals were now much better. A few days after injection it was clear that Pasteur had triumphed. All the vaccinated animals were well, whereas twenty-one out of the twenty-four non-vaccinated sheep were already dead; two of the surviving ones died in front of spectators who had come to see the spectacle. The last remaining non-vaccinated sheep died later the same day. The four non-vaccinated cows were all swollen with anthrax and had fever; and the non-vaccinated goat was dead of anthrax. Over the next few days only one of the vaccinated animals died and it was a sheep that died as a result of pregnancy complications: its blood did not contain anthrax bacilli.

Pasteur had quietened his opponents and became a national hero in France. Use of his attenuated anthrax vaccine spread all over France and abroad; France's incidence of anthrax in cattle fell almost fifteen-fold. His achievements had been manifold (Chapter 11): he had discovered molecular asymmetry, proved the microbe theory of fermentation, refuted the notion of spontaneous generation, confirmed the germ theory of disease, and now developed vaccines against chicken cholera and anthrax. Even more was to come. Pasteur developed a vaccine against swine erysipelas, this time attenuating the microbe by passing it through laboratory rabbits, a process that made the microbe less virulent for pigs and provided protection against subsequent infection. Then came his celebrated vaccine against rabies.

At that time rabies in humans was not particularly common in France: there were only a few hundred deaths every year as a result of the disease. However, the horrific symptoms of rabies were well known, and treatments that were used to try and cure it were drastic. Symptoms usually took some time, frequently a month or more, to appear after the victim was bitten by a rabid animal (usually a dog or wolf). The disease affected the nervous system, producing restlessness, viciousness, seizures and paralysis; victims often exhibited a fear of the sight of water and so rabies was also called hydrophobia. Death occurred within a week after symptoms appeared. It was not uncommon for human rabies victims to be deliberately killed to put them out of their misery, and the wounds of people freshly bitten by mad dogs infected with rabies were sometimes cauterized using red-hot iron in an effort to prevent the rabies from taking its grip on the body.

Although rabies was known to be contagious and saliva of infected animals was known to contain the infectious microbe that caused it, nobody had identified the microbe. Pasteur spent a good deal of time searching for the rabies germ, but he failed to find one. We now know why: the cause of rabies is a virus, not a bacterium, and since viruses are much smaller than bacteria, they would not have been detectable using the microscopes available in the nineteenth century. The same was true for smallpox: it was caused by a virus and so no-one had been able to identify any infectious organism under the microscope. However, Pasteur was not deterred by the fact that he could not find a rabies microbe: the disease clearly was infectious and showed all the features of a microbial disease. When saliva of infected dogs or humans was injected into rabbits it produced the symptoms of rabies, and so the invisible microbe was obviously present in saliva. The infectious agent was also present in brain and spinal cord matter, since these also could be used to induce rabies in experimental animals.

Pasteur discovered that he could not cultivate the invisible rabies virus in his usual broths. We now know the reason for this: viruses cannot multiply on their own, but instead they require living cells of the body, within which they multiply. Today viruses are often grown in laboratories in flasks containing isolated living cells taken from a human or other animal. Although the inability to grow rabies virus in broth was a hindrance to Pasteur, it did not stop him from seeking a possible rabies vaccine. Instead, he developed a method for growing the virus in live animals. Some of the work was distressing to Pasteur because it involved a certain amount of animal suffering, but he had a reputation for being extremely kind and gentle to his animals, to the extent that he sometimes delayed experiments because of his concern.

Anti-vivisectionists of the day had already harassed Pasteur, but he firmly believed that the good that came from his work far outweighed the negative aspects of his animal experiments. The myriads of lives that have been saved by Pasteur's discoveries confirm his belief and, as Pasteur himself said, many animals have benefited from his work as a result of its applications to agriculture and veterinary medicine.

Pasteur investigated the possibility that the rabies virus could be attenuated for use as a protective vaccine in a similar way that he had developed attenuated forms of chicken cholera, anthrax and swine erysipelas microbes. When spinal cords of rabid rabbits were hung up and left to dry for a few weeks inside sterile flasks, the virus in the spinal cords was no longer infectious to dogs: it had become attenuated. A series of inoculations of this dried spinal cord material was able to protect dogs from rabies: the attenuated virus served as an efficient rabies vaccine. Because the symptoms of rabies took some time before they appeared following a bite from a rabid animal, Pasteur wondered if the vaccine might not only protect dogs from infection, but also if it would prevent symptoms from appearing in a dog that was already infected with the disease. He vaccinated a series of dogs some days after they had contracted rabies, but before their symptoms developed, and discovered that the attenuated virus could, indeed, prevent rabies from taking its grip on the animals.

The step from dogs to humans was not an easy one for Pasteur to make because there were uncertainties as to whether the method of vaccination so carefully worked out in dogs would apply equally to humans, and Pasteur was well aware of the horrors of the symptoms of rabies, and he obviously feared the possibility that an unsuccessful vaccine preparation might induce the disease rather than prevent it. However, in 1885 a nine-year-old boy called Joseph Meister, who had been badly bitten two days earlier by a rabid dog, was taken to Pasteur's laboratory by his mother. The young boy had fourteen bites on his body and his doctor gave him no hope of survival: he would inevitably show the terrible symptoms of rabies and die. Pasteur was greatly moved by Meister's situation and decided to inoculate him with attenuated rabies virus obtained from laboratory rabbits. The boy survived without symptoms and eventually became a gatekeeper at the Pasteur Institute in Paris. He lived until 1940, when he died in an incident in which German soldiers of Hitler's Nazi regime tried to force him to open the entrance to the crypt where Pasteur was buried: rather than allow the soldiers access to 'The Master's' tomb, Meister committed suicide.

Meister's vaccination was followed by other successes, and although there was a good deal of controversy over the use of Pasteur's rabies vaccine in humans (some even branded Pasteur as a murderer), there is no doubt that many lives were saved by it and thousands of people were indebted to Pasteur. The rabies vaccine was an extension of the previous three vaccines that Pasteur developed, and it brought home the clear message that vaccines were attainable for a number of diseases, that they were safe, and that they provided powerful protection from disease.

People of the twentieth century have seen the development of vaccines against many diseases; the methods used to produce these vaccines were based on Pasteur's own procedures. The chief aim has been to produce an attenuated form of the disease agent, or some component of it, that induces a protective immune response in inoculated individuals. Once a person has been exposed to the vaccine, his or her immune system 'remembers' the encounter with the vaccine and subsequently fights off anything that resembles the vaccine, such as the real pathogen itself. The early studies of Jenner and Pasteur have led to an extensive understanding of the immune system and how it destroys invading microbes and viruses. Organ transplantation has benefited from this knowledge of the immune system, since tissue rejection involves destruction of cells of transplanted organs by the patient's immune system. The success of matching donor tissues with organ recipients and of using immunosuppressants to encourage transplanted organs to 'take' have depended on a detailed knowledge of the immune system.

By the early part of the twentieth century it was clear that infectious diseases could be caused by bacteria and viruses. Another group of parasitic micro-organisms, the protozoa, had also been shown to cause diseases, especially in tropical countries, and some of these diseases were more complicated than many of the bacterial and viral diseases that were transmitted directly from one person to another. Many protozoal diseases were shown to be transmitted from person to person by a biting insect. The first of these diseases to be discovered was malaria, which is transmitted by mosquitoes.

13

Malaria's cunning seeds

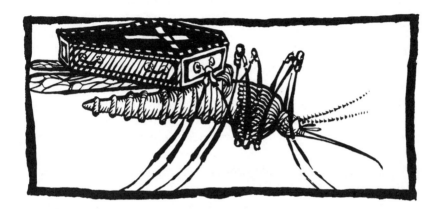

During the US Civil War more than half of all the soldiers became infected with malaria. Indeed, throughout history malaria has killed more soldiers than have died as a result of war, and it has almost completely wiped out some human populations. Yet today malaria no longer is a threat in the USA and it has been eliminated from scores of other nations in which it was once rife. This purge of one of humanity's major health problems from many countries is due in large measure to a British physician, Sir Ronald Ross (1857–1932), who demonstrated that malaria is carried by mosquitoes. Ross's great achievement led to public health measures, particularly spraying of the insecticide, DDT, for controlling the mosquitoes that carry malaria.

The scale of the importance of Ross's discovery and the subsequent efforts aimed at reducing mosquito-borne malaria transmission are evident when one considers that malaria is still an enormous problem in more than a hundred developing countries, where insufficient control measures have been taken. In 1992, the World Health Organisation announced that almost three hundred million people were infected with malaria world-wide, and that more than a million people die every year of the disease. Malaria is endemic in Central and South America, the Middle East and Asia, and in countries bordering the Mediterranean Sea. About two thousand million people (nearly half of the world's population) live in areas where malaria-carrying mosquitoes thrive, and in some parts of Africa almost a quarter of the children under five die from malaria every year.

The symptoms of malaria include alternate bouts of high temperatures and chills, anaemia, jaundice, and swelling of the spleen and liver. In many cases, the disease reaches the brain, leading to loss of body control, loss of consciousness and abnormal behaviour. Death usually occurs once the disease has spread to the brain.

Our understanding of malaria has come a long way since Hippocrates (460–370 BC), the Father of Medicine, described its symptoms. Ronald Ross's discovery that mosquitoes carry the disease was undoubtedly the most important step forward ever made in the study of malaria, because it explained how malaria was spread amongst human populations and paved the way for possible means of control of the disease. Ross received the Nobel Prize for Physiology and Medicine in 1902, in recognition of his work. If, in the future, the knowledge gained from his discovery is eventually applied properly to eradicate malaria from all nations of the Earth, his contribution will be seen as a monumental landmark in the fight to eliminate one of the world's most devastating diseases.

Ronald Ross and the malarial life cycle

The ancient Greeks and Romans were familiar with malaria and its characteristic intermittent bouts of sweating and shivering. The Roman scholar, Marcus Terentius Varro (116–27 BC), advised anyone who intended to build a farmhouse to avoid swampy land, which, he believed, might be a source of malarial fever. 'Certain minute animals, invisible to the eye, breed there, reach the body by way of the mouth and nose and cause diseases', he said. Varro's belief that malaria was caused by 'minute animals' was shown to be correct two thousand years later. Pliny the Elder (AD 23–79), a Roman scientist, even suggested various remedies for malaria, one of which was to wear a 'lucky charm' consisting of a 'green lizard enclosed in a vessel just large enough to receive it.'

The idea eventually arose that malaria was caused by evil emanations, or 'miasma', from swamps and marshes: the word, 'malaria' means 'bad air' in Italian. The exact nature of these bad airs was unclear, but they were believed to rise into the atmosphere and cause malaria when they were inhaled or eaten as food contaminants.

Towards the end of the nineteenth century the germ theory of disease – the idea that many human infections were caused by microscopic living organisms – was gaining widespread acceptance (Chapter 12). Some scientists also believed that malaria might similarly be caused by a microscopic organism rather than by the mysterious miasma. A crucial discovery was made in 1880 by the French physician, Charles Louis Alphonse Laveran (1845–1922), who was working as a surgeon with the French army in Algeria. Laveran examined the blood of malaria patients and noticed a microscopic organism inside their red blood corpuscles. He believed (correctly) that this organism, which was later called a **plasmodium**, was the cause of malaria, supporting the idea that malaria was due to a germ.

The plasmodium belongs to a group of one-cell organisms called **protozoa**, which are biologically more complicated than bacteria. Laveran's discovery was the first time that anyone had found a protozoan that caused a human disease. Laveran was later awarded the 1907 Nobel Prize in Physiology and Medicine for his work on malaria and other protozoan diseases. Subsequently, other protozoa were found to cause some of the most horrific diseases of humankind, particularly tropical diseases such as African sleeping sickness, leishmaniasis, amoebic dysentery and Chagas' disease. These diseases are still rife today in countries of the developing world and cause untold human suffering.

Whilst Laveran's discovery was accepted by many scientists, nobody knew how plasmodia were transmitted to humans. Many believed that plasmodia were present in domestic water and that a person caught malaria by drinking plasmodia-contaminated water. An English physician, Albert King (1841–1914), had another idea about malarial transmission. King published an article in 1883 in which he stated nineteen reasons why he thought malaria was transmitted to humans by mosquito bites. In particular, he said, malaria was common in swampy, marshy or jungle regions, where mosquitoes were known to thrive, and malaria was known to be especially easy to contract during the night, when mosquitoes fed on human blood.

King's theory was all the more credible because a Scottish physician, Patrick Manson (1844–1922), had shown that mosquitoes carried another human disease, **filariasis** (elephantiasis), which is caused by a parasitic worm. Manson also believed King's theory that mosquitoes carried malarial plasmodia, although he thought that mosquitoes simply transferred plasmodia from marshes or swamps to domestic drinking water, and that the protozoan gained entry into the human body in drinking water rather than during the bite of the mosquito. (Ross later disproved this by drinking large volumes of water contaminated by mosquitoes, which left him without any ill effects.)

It was at this point that Ronald Ross began to study malaria in depth. Ross had joined the Indian Medical Service in 1881 as an army surgeon and left for India in the same year. He became interested in mosquitoes in 1883 whilst he was in Bangalore, particularly because he was frequently bitten by them. He studied the different species of mosquito and made distinctions between their appearance. In 1894 Ross took a holiday in England, where he wrote an essay in which he described arguments against the idea that malaria was caused by 'bad airs'. The essay won him the Parkes Memorial Prize of 1895, which was given to the author of the year's best monograph on malaria. Ross demolished the 'bad air' theory using clear scientific arguments.

While he was in England Ross visited Patrick Manson and the

two scientists discussed malaria. Although Ross had observed blood from malaria patients for several years, he never saw Laveran's plasmodia inside their red blood corpuscles. Manson, however, convinced him that Laveran was correct by taking some blood from a malaria patient in a London hospital and showing Ross the blood under a microscope. Ross was left with no doubts as to the existence of malarial plasmodia in sufferers of the disease.

The visit to Manson also convinced Ross that mosquitoes were responsible for transmitting malaria. However, nobody had any direct evidence for the mosquito theory. Ross realised that if there was a link between malaria and mosquitoes, then plasmodia should be detectable in mosquitoes. He invented a portable microscope to take back with him to India and set himself the task of showing that plasmodia could be found in mosquitoes fed on malaria-infected human patients.

The apparently simple task Ross had set out to achieve turned out to be much more difficult than he had imagined. At first he had difficulty in finding malaria patients to study. Then, when he did find the patients with plasmodia in their bloodstream, he encountered further problems. For example, some patients ran away after he pricked their fingers to collect a drop of blood; captured mosquitoes frequently died before they could feed on a patient; and many mosquitoes simply would not bite a patient. Eventually, he found that mosquitoes were more likely to bite if the patient's bed and mosquito net were dampened with water. An even better method was to place the mosquito inside a test-tube and push the open end of the tube against a patient's skin. By the end of 1896 Ross had bred as many different species of mosquito as he could capture and he had seen hundreds of patients with plasmodia in their blood cells. However, despite his success at feeding mosquitoes on these patients, he had not seen a single patient-fed mosquito containing plasmodia.

Ross's studies of malaria were often considered to be a nuisance by senior officers in the army, who sometimes held back his research by giving him tasks that distracted him from malaria, or by posting him to areas where he could not easily work on malaria. Little did they realise how important his malarial investigations were. Towards the end of 1896, Ross decided to take two months' leave from the army so that he could continue his malaria research unhindered. This he did: instead of going on a pleasant holiday he went to a highly malarious region of India so that he could gain access to patients and mosquitoes and continue his research. Even there, however, he failed to find plasmodia in mosquitoes, although he suffered an attack of malaria himself.

Following a bout of cholera which nearly killed him, Ross returned to the army in India and carried on his search for plasmodia in mosquitoes. He arranged for his assistants to search for mosquitoes, which he then fed on patients. These mosquitoes were then dissected under a microscope and examined for the presence of plasmodia. The amount of painstaking work involved was enormous: every square micrometre (one-thousandth of a millimetre, less than a ten-thousandth of an inch) of each mosquito was scanned. The process took hours for just one mosquito to be examined and thousands of the insects were investigated. With little success, Ross was still convinced that the plasmodia were transmitted by mosquitoes. Most people would have given up the search by now, but not Ross: so certain was he of the truth of the mosquito theory of malaria transmission. 'The things are there, and must be found! It is simply a matter of hard work', he wrote to Manson.

On 16 August 1897, Ross's assistant brought him twelve **Anopheles** mosquitoes, a type that he had seen infrequently until then. At the time, Ross called these mosquitoes 'speckled', since he was not trained in entomology and did not know to what species they belonged. He fed them on malaria patients and left them for several days for any possible plasmodia to establish themselves in the insects. However, by 20 August only three mosquitoes remained – the others had died. One of the three left also died, and when Ross examined it under the microscope he found no plasmodia there. He dissected and scanned one of the remaining two mosquitoes: it contained no plasmodia.

By this time, Ross was tired and beginning to doubt his theory, and he wondered whether or not the last mosquito was worth dissecting and examining. He decided to go ahead, and placed the last mosquito on a microscope slide, dissected it, and spent half an hour searching, in vain, for plasmodia. Then he focussed his microscope on the mosquito's stomach contents and, to his great excitement, there he saw not one, but several pigmented cells that resembled the plasmodia he had so frequently seen in malaria patients. His perseverance and diligence were rewarded with a great discovery – the presence of malarial plasmodia in the stomach of a mosquito.

Ross realised the significance of his discovery, although as a careful and objective scientist he knew that he had yet to prove beyond reasonable doubt that the pigmented cells in the mosquito stomach were, indeed, malarial organisms and not some other mosquito parasite that only resembled plasmodia. It would be especially important to demonstrate that a plasmodia-infected mosquito can cause malaria when it bites a non-infected person or animal.

Ronald Ross was not only a physician and scientist: he also wrote poems, fiction, articles on mathematics and chamber music. Soon after finding plasmodia in mosquito stomachs he wrote a poem describing his discovery:

> This day relenting God
> Hath placed within my hand
> A wondrous thing; and God
> Be praised. At His command,
>
> Seeking His secret deeds
> With tears and toiling breath,
> I find thy cunning seeds,
> O million-murdering Death.
>
> I know this little thing
> A myriad men may save.
> O Death, where is thy sting?
> Thy Victory, O Grave?

The finding of what appeared to be malarial plasmodia in *Anopheles* mosquitoes was the key to understanding malarial transmission. Ross now realised why he had failed for so long to find plasmodia in mosquitoes fed on human patients: he had been using the wrong species of mosquito. Now he could focus on *Anopheles* mosquitoes, the real plasmodia carriers. Soon afterwards, he again found plasmodia in the stomachs of these mosquitoes and this strengthened his opinion that he had, indeed, found the link between the mosquito and human malaria.

Ross was ready to take his great discovery to its inevitable conclusion, that is, to demonstrate that plasmodia-infected mosquitoes cause malaria when they bite a person. Then, in September 1897 his research was cut short by the bureaucracy and obstructiveness of his superiors: he was ordered to the front on active service. Despite his protests, he had to stop his malaria work. The significance of his research was not appreciated at all by the military authorities, even though soldiers were fully aware of the horrifying nature of malaria. However, Manson and other scientists protested strongly to the army authorities and three months later Ross was allowed to spend six months in Calcutta, India, on special service. There he was able to complete his important research.

In Calcutta, Ross had difficulty in finding human malaria patients who were willing to provide him with the much needed blood samples, but he did make the highly significant observation that the local birds (specifically, sparrows) suffered from malaria. This provided him with an excellent model system for studying malaria transmission. Within a few months he traced the whole of the life cycle of plasmodia in birds. Birds infected with malaria transferred malaria to mosquitoes, and mosquitoes infected with plasmodia transmitted malaria to birds. He demonstrated that plasmodia pass from the mosquito's stomach into its salivary glands and are transferred to birds in the saliva of mosquitoes when they feed on blood during a bite. Although the species of mosquito that transmits bird malaria differs from that which transmits human malaria, and bird malaria cannot be transmitted to humans, the life cycles of human and bird malaria are very similar.

In April 1898, Ross wrote again to Manson, 'I consider the mosquito theory to be absolutely proved.' Very soon afterwards, a group of Italian scientists extended his findings by showing that mosquitoes transmitted plasmodia by biting humans in the same way that they caused bird malaria.

The life cycle of the malarial plasmodium in humans (Figure

21) has been somewhat refined since Ross's time, but all his results and his scientific interpretations of them remain unchanged today. Knowledge of the life cycle has directed research and public health programmes aimed at eliminating malaria from the face of the Earth. Research programmes today are setting their sights towards development of a possible vaccine for malaria. Such a vaccine would block the life cycle at one or more points and prevent continued transmission of the plasmodium from humans to mosquitoes and back again. Drugs are available that are effective at treating malaria, but a major problem exists that some strains of the plasmodium are becoming resistant and consequently cannot be destroyed by these drugs.

Unfortunately, malaria is still a major problem despite Ross's efforts. Ross believed that knowledge of the plasmodial life cycle could be exploited to reduce drastically the incidence of malaria, and he proposed public health programmes aimed at destroying mosquitoes and giving patients the anti-malarial drug, quinine. The British authorities were very slow to listen to him, however, even though there were tremendous human and commercial benefits to be gained from reducing malaria incidences in countries of the British Empire. The Americans were quicker to respond. In 1904 Ross was invited by the US public health authorities to visit the Panama Canal whilst it was being built and, using his proposed malaria control measures, the disease was eliminated from the area. Those building the Panama Canal were also freed from yellow fever, another disease that is carried by mosquitoes.

Following his Nobel Prize, Ross was knighted for his work. Ross's research signalled the beginning of the end for malaria in many parts of the world; however, the incidence of malaria is still very high in developing countries. Ross himself was very disappointed that his ideas for malaria eradication were not taken seriously enough in many places, and he would unquestionably be astounded and shocked if he knew that the disease was still rife today, almost a century after his great discovery. Heroic efforts are being made to eliminate malaria, but far more needs to be done. Ronald Ross pointed the way.

In 1917, two decades after he first discovered plasmodia inside

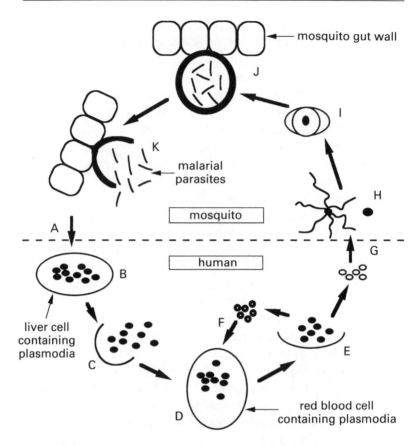

Figure 21.   Life cycle of the malarial plasmodium. The top
half of the diagram shows the parasite's development in the
mosquito; the bottom half shows its development in humans.
Plasmodia are injected into human blood during the bite of a
mosquito (A). These reach the liver and invade liver cells,
where they divide and develop (B), eventually being released
(C) as forms that invade red blood corpuscles. Plasmodia divide
inside red blood cells (D) and are eventually released (E). At
this stage, the plasmodia can either re-invade red blood cells
(F), or they can enter mosquitoes during another bite (G). In
the insect, plasmodia undergo a sexual process involving male
and female plasmodia (H, I) and bore through the gut wall of
the mosquito, where they produce forms that are infectious to
humans (J). These burst out from the gut wall of the mosquito
(K) and migrate to the salivary glands. They re-enter a human
victim in mosquito saliva when the insect feeds on a human.

the stomachs of *Anopheles* mosquitoes, Ross wrote another poem:

> Now twenty years ago
> This day we found the thing;
> With science and with skill
> We found; then came the sting –
> What we with endless labour won
> The thick world scorned;
> Not worth a word today –
> Not worth remembering.

As we approach the twenty-first century, we can but hope that scientific research, public health measures and governments of the world will exploit the work of Sir Ronald Ross with enough zest to make sure that malaria, like smallpox, becomes a disease of the past.

14
Penicillin from pure pursuits

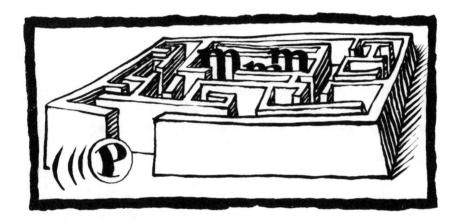

T he idea that micro-organisms caused some diseases was
widely accepted by the beginning of the twentieth century
and many scientists were considering the possibility that these
diseases might be treated using chemical substances that kill
microbes but are not harmful to humans. Coupled with vaccines
for preventing various diseases, these chemotherapeutic agents
would provide a powerful tool in the fight against infectious
diseases.

The first drug used regularly to treat a human bacterial infection was **salvarsan**, also called Compound 606, which was discovered by the great German scientist, Paul Ehrlich (1854–1915).
Ehrlich can be considered to be the founder of chemotherapy.
He originally worked with certain dyes and demonstrated their

ability to selectively stain bacterial cells but not human cells. If some dyes stain bacteria but not human cells, he thought, might it not be possible to develop chemical drugs that also selectively kill bacteria and leave their human hosts unharmed? His idea was that it may be possible to make a kind of chemical 'magic bullet', which would seek out invading microbes in the body and kill them.

In 1905, Ehrlich began a research programme in which thousands of chemical compounds were screened for their effects on various diseases. Salvarsan, an arsenic-containing chemical, was particularly effective at curing syphilis and soon became widely used in its treatment. Many of today's drugs that are used to treat sleeping sickness and other tropical diseases caused by protozoan microbes were derived from Ehrlich's work. Indeed, drugs against protozoan diseases were available before drugs were produced to treat most bacterial diseases. Paradoxically, protozoan diseases today are highly neglected by the pharmaceutical industries, even though these diseases cause misery and devastation to hundreds of millions of people in developing countries.

Salvarsan was the only true 'magic bullet' available for any bacterial disease for many years. Then, in the 1930s the German biochemist, Gerhard Domagk (1895–1964), applied Ehrlich's principles of chemotherapy to infectious diseases by testing a series of newly produced dyes as possible drugs against bacterial infections in mice. Domagk found that one dye, which was called **Prontosil**, was a highly effective antibacterial. This was a major breakthrough because Prontosil was able to attack bacteria belonging to the class called streptococci, which are found in some important human infections, such as scarlet fever.

Interestingly, Prontosil has no effect on bacteria outside of animals: the drug does not kill isolated bacteria in a test-tube. So if Domagk had not carried out his tests on infected mice, he would never have discovered Prontosil. The reason for this is that Prontosil is converted to another substance in the body and it is this substance that attacks the bacteria. Prontosil was the first ever **sulpha drug**, and many more similar drugs were subsequently made as a result of Domagk's discovery; some are still very useful for certain bacterial infections. The discovery of Prontosil demonstrated that chemical drugs could, indeed, be used against a wide variety of bacterial infections, and it had a strong influence on the subsequent applications of penicillin as an antibiotic.

Sulpha drugs had their wonderful applications, but they also had their drawbacks. They sometimes produced undesirable side-effects and there were still many disease-producing bacteria that were unaffected by these drugs. Unknown to the world, the most important chemotherapeutic agent yet available for the treatment of bacterial diseases had already been discovered, in 1928, well before sulpha drugs were available. This forgotten drug was **penicillin**, and its significance as a disease-curing substance was not appreciated until 1940. The story of the discovery of penicillin is truly remarkable and it is one of the most extraordinary examples of serendipity in the history of science.

Penicillin was discovered by the Scottish bacteriologist, Sir Alexander Fleming (1881–1955), but it was the Australian pathologist, Sir Howard Florey (1898–1968), the German biochemist, Sir Ernst Chain (1906–1979), and their colleagues in Oxford University who, twelve years later, resurrected Fleming's penicillin and gave the world an antibacterial drug with a versatility and power never before seen.

There is strong evidence that penicillin had been encountered by numerous scientists even before Fleming's discovery. For example, in 1871 Joseph Lister, who developed antiseptic surgery as a result of hearing about Pasteur's work on microbes, noticed that one of his bacterial cultures failed to grow when it was accidentally contaminated with *Penicillium* mould. Lister carried out some experiments with the mould in order to assess its use as an antiseptic agent, but his results were not conclusive. However, a hospital record made in 1884 indicates that he used *Penicillium* mould in the successful treatment of an abscess on one of his patients. Nevertheless, Lister seems not to have pursued with any further research on the mould, and the real history of penicillin begins with Sir Alexander Fleming.

Fleming and penicillin

Alexander Fleming was born in Scotland in 1881, the son of a farmer. When he was sixteen he worked for a shipping company as a clerk, but his older brother believed that he was suited for a more intellectually demanding career. When a relative died, Alexander was left some money in the will, and he used this to support a medical career. He moved to London, gained the necessary qualifications, by private tuition, and obtained a place to study medicine at St Mary's Hospital in London.

Fleming worked in the Inoculation Department, where the head of the department, Sir Almroth Wright (1861–1946), had research interests that included vaccines and the immune system. For some years Fleming worked as a researcher and at the same time he practised medicine, but after the First World War he concentrated his efforts solely on research. He had a reputation as a shy man who found conversation somewhat difficult, but he was nevertheless highly respected and was known to have a keen eye for detail.

In 1922 Fleming discovered **lysozyme**, a protein that breaks down bacteria. He was suffering from a bout of common cold, and allowed a drop of mucus from his nose to fall onto the surface of nutrient medium in a bacterial culture dish. Some days later he discovered that bacteria were not growing in a region around the mucus, but they were growing some distance away from it. It appeared that the mucus contained a substance that was lysing the bacteria. Fleming called this substance lysozyme and he produced some evidence that it was present in various living organisms and body fluids, including tears, saliva, mucus, egg white and plants. However, whilst lysozyme was able to kill many harmless species of bacteria, it did not kill most bacteria that cause human diseases. Lysozyme eventually became a widely studied protein and has served as a model enzyme (a protein that catalyses a chemical reaction) in biochemical studies.

The discovery of lysozyme was very similar to Fleming's later discovery of penicillin, and it is likely that the discovery of penicillin was made somewhat easier by Fleming's experience with lysozyme.

In 1928 Fleming became a professor of bacteriology at St Mary's Hospital. His main research topic involved studies of the class of bacteria called staphylococci, which cause problems such as boils, septic spots and some forms of pneumonia. Fleming grew staphylococci from the boils of hospital patients on the surface of solid nutrient jelly in culture dishes. He investigated these bacteria and noticed the various colours of colonies formed as a result of their multiplication. Fleming was a diligent and conscientious worker. His laboratory was very small, measuring only 3.6 metres by 3 metres (12 feet by 10 feet), and it was consequently extremely cluttered.

The events surrounding the discovery of penicillin by Fleming are not entirely clear, largely because the magnitude of the discovery was not appreciated at the time. Later, when penicillin became a major antibacterial drug, scientists who worked with Fleming tried to recall the exact events that took place, but some of these accounts varied and we may never know the precise details. However, the following story is probably near to the truth.

In the summer of 1928, Fleming went on holiday, but before he left he collected his bacterial culture dishes together and piled them up at one end of his laboratory bench so that he could examine them when he returned from his vacation.

After the summer break, in September 1928, Fleming decided to look carefully at the pile of culture dishes he had left over the holidays: he always examined each plate carefully and gleaned the maximum amount of information from them. The bacterial colonies on every dish provided some information. He left the plates on the bench and carried on with the rest of his work. Later on, a colleague came to see him and Fleming pointed out the many plates that he had earlier examined and began showing the colleague the various bacterial colonies on the plates.

It was then that Fleming noticed something rather unusual and interesting on one of the plates. The surface of the culture jelly contained many staphylococcal colonies dotted here and there and it also had a blob of fluffy fungal mould at one edge of it. But, more interestingly, the area immediately around the mould contained very few bacterial colonies, and the ones that were there looked transparent as if the staphylococci in them were dying or dead. Much further away from the mould the bacteria looked perfectly happy. This historic plate was probably inches away from destruction: it is thought by some to have been near the top of a pile of plates that were being discarded, in a tray of bleach. Had it been further down the pile, it might well have been destroyed forever!

This discovery was reminiscent of Fleming's earlier discovery of lysozyme, when he had noticed clear areas with no bacterial colonies surrounding a drop of nasal mucus. Here, the mould appeared to be doing a similar thing to the lysozyme in the nasal mucus. Fleming immediately recognised the situation, and at first he thought he must have discovered a type of lysozyme that was produced by a fungus. He must have been somewhat excited by the discovery because he had the culture dish photographed and he showed it to many other scientists in the department, although the degree of interest from them was generally rather mild. He also preserved the plate and kept it for years; it is now in the British Museum in London.

Fleming realised that there was an interesting phenomenon to be studied, so he removed some of the fungal mould, transferred it to a flask of liquid nutrient broth and grew it in large amounts. He took some of the liquid in which the mould was grown and found that it was able to inhibit growth of staphylococci in culture dishes, just as the area around the mould inhibited bacterial growth in the original dish. Evidently, the mould was secreting a substance that was harmful to staphylococci. When Fleming examined other bacteria for their sensitivity to the mould juice he discovered that some bacteria were not affected, whilst the growth of other types of bacteria was hindered. Unlike lysozyme, however, the mould juice seemed to be able to attack many types of bacteria that cause human diseases, and Fleming realised that the substance produced by the mould was not a type of lysozyme.

Over the next few months Fleming excitedly investigated the antibacterial substance in mould juice, which he called 'penicillin' because the mould producing it was identified as being a species of the fungus, *Penicillium*. He also examined thirteen different species of mould, including eight different types of *Penicillium*, for production of penicillin and only one of them – the original type of mould in which penicillin was found – produced any penicillin. This shows just how lucky Fleming's discovery had been: penicillin-producing moulds are rarely found as contaminants on bacterial culture dishes.

He carried out some basic studies on the properties of penicillin and found that it was quite powerful: mould juice diluted one part in eight hundred still was able to attack bacteria. Penicillin did not have any effects on human white blood cells, and when Fleming injected large quantities of the substance into rabbits or mice, there were no toxic effects. Penicillin was highly damaging to disease-causing bacteria, yet it appeared to be harmless to animals and to human white blood cells. Fleming believed that penicillin might be a useful antiseptic agent, like Lister's carbolic acid, that could be used on the skin or wounds; and it seemed to be safer than carbolic acid, which did kill human white blood cells.

One of the members of Fleming's research team actually ate some of the *Penicillium* mould and suffered no problems, and he also used the mould juice containing penicillin to irrigate an infection he had in his nose. The infection did not clear up, but the penicillin did no harm. Another of Fleming's colleagues flushed an infected eye with the mould juice and this time he was cured. Fleming also used penicillin as an antiseptic, instead of carbolic acid, on the wounds of a few patients at St Mary's Hospital: results were good in some cases but generally they were not spectacular. Fleming gave up the idea of using penicillin as an alternative antiseptic, at least until more of it could be obtained in a state of greater purity than it existed in mould juice.

Two of Fleming's co-workers decided to try and purify penicillin from other components in mould juice, so that it could be obtained free from impurities in larger amounts. Unfortunately, penicillin seemed to be unstable and the scientists had a great deal of difficulty in purifying it. They gave up eventually, and little further effort was made in Fleming's laboratory to purify penicillin. In 1934 another of Fleming's colleagues tried again but with the same lack of success. In 1932, another scientist, at the London School of Hygiene and Tropical Medicine, had also tried to purify penicillin from culture broth containing *Penicillium* mould that had been supplied by Fleming. Again, the attempt was in vain: penicillin was just too unstable and could not be obtained from mould juice in reasonable quantities.

Fleming continued to use penicillin to get rid of unwanted bacteria from his cultures: if a type of bacterium was not sensitive to penicillin, any contaminating bacteria that were sensitive could be kept out of the cultures by adding mould juice to the culture. Even though he had published reports of his discovery, most other scientists considered the discovery not to be particularly remarkable and few other laboratories took up the work, except to rid their cultures of unwanted bacteria. Perhaps strangest of all, Fleming did not carry out any experiments involving treatment of bacteria-infected animals with penicillin to see if they were cured. He knew that penicillin was not toxic, and he knew it killed disease-causing bacteria, but he never carried out that crucial experiment. It appears that one reason for this was that the other work with penicillin, especially its instability and the slowness with which it killed bacteria in his experiments, made him less enthusiastic about its possible use as a chemotherapeutic drug. Another possibility is that the department in which he was working was more interested in fighting diseases with vaccines than with drugs and so the degree of interest in penicillin as a drug was minimal.

Whatever the reasons, the world had to wait more than ten years after penicillin's discovery before it was used as a wonder drug. Fleming had made the crucial and serendipitous step of discovering penicillin, the curious laboratory antibacterial, and his work and that of others on its properties had paved the way for the 'rediscovery' of penicillin, the great antibiotic drug of the twentieth century. It was in Oxford University at the beginning of the Second World War that penicillin's potential was eventually realised, in a laboratory headed by Professor Howard Florey (1898–1968).

Oxford and the realisation of penicillin's potential

Florey became a professor of pathology in the Sir William Dunn School of Pathology in Oxford in 1935. His research interests were particularly concentrated on lysozyme; he started working on lysozyme in 1929, some years after Fleming had discovered it. Indeed, Florey had collaborated with Fleming on one particular study of lysozyme. He wanted to continue his studies of the biochemical properties of lysozyme and began to search for a suitable research biochemist to join him. The person eventually employed in this role was Ernst Chain, a brilliant biochemist who had come to Britain a few years earlier from Berlin: Chain was Jewish and had fled Hitler's persecution of Jews in Germany.

Chain studied lysozyme in Oxford for some time, but eventually he and Florey decided to extend their research to other antibacterial agents. In 1938 Chain carried out a search of scientific papers that had been published on antibacterial substances that were produced by other micro-organisms, such as fungi. One of the papers he found was the one on the discovery and properties of penicillin that had been published by Alexander Fleming in 1929. Chain was immediately interested in this paper and thought at first that penicillin was similar to lysozyme, just as Fleming himself had done.

Penicillin was chosen as one of several antibacterial substances that Florey and Chain would investigate. By coincidence, another researcher in the Sir William Dunn School of Pathology had previously obtained some *Penicillium* mould from Fleming and was using it in the same way as Fleming, that is, to remove unwanted bacteria from his cultures. Chain procured some of the mould and began to investigate penicillin with a view to purifying it and examining its role as an antibacterial substance. One of his colleagues, Norman Heatley (b. 1911), played a major role in getting the project off the ground.

Both Florey and Chain later said that their motivation behind the work on penicillin was purely to understand it on a scientific basis, and that they were not particularly concerned about the possibility that it might be a useful drug. Florey said, 'People sometimes think that I and the others worked on penicillin because we were interested in suffering humanity – I don't think it ever crossed our minds about suffering humanity; this was an interesting scientific exercise.' Chain agreed, 'The only reason that motivated me was scientific interest. That penicillin could have a practical use in medicine did not enter our minds when we started to work on it.' Nevertheless, as Florey, Chain and their colleagues found more about the remarkable properties of penicillin, it became clear that they were working on a wonder drug, and their thoughts did eventually turn to suffering humanity.

Penicillin was a stimulating challenge to Chain, who, as a bio-chemist, was keen to try and purify it. Chain was particularly interested in enzyme proteins, and initially he thought that peni-cillin was an enzyme. It soon turned out, however, that penicillin was much smaller than a protein, perhaps to the disappointment of Chain. As progress was made, Florey decided to go full steam ahead in the study of penicillin. Florey was somewhat weary that British funding bodies either rejected his grant applications or only gave him small sums of money for his research, so he applied to the Rockefeller Foundation in the USA for substantial funding and was successful. The Rockefeller Foundation provided most of the initial research funds for Florey and Chain's work on penicillin.

By 1940 Florey, Chain, Heatley and their co-workers had pro-gressed tremendously in their study of penicillin. They had an improved method for measuring its potency as an antibacterial substance; they found that greater amounts of penicillin could be obtained if the mould was grown in shallow culture vessels shaped like bed-pans; and they had preparations of penicillin of increased purity. Indeed, Chain had a small amount of a penicillin prep-aration that was much more powerful than any previously avail-able: diluted one in a million, Chain's penicillin still killed bacteria. It was substantially more powerful even than any of the sulpha drugs that were being used at the time to treat human infections.

Chain had a colleague inject some of his penicillin preparation into two mice and showed that it was not poisonous to the mice;

this confirmed the results obtained by Fleming twelve years earlier. Florey decided to carry out some more detailed studies with animals in order to find out what happened to penicillin when it entered the body. He discovered that penicillin could not be taken by mouth because it was destroyed in the stomach; but it did remain in the body some time after it was injected into the bloodstream. Eventually any penicillin in the body was excreted in urine. Nowadays, new variants of penicillin are available that can be taken by mouth: they are modified versions of the original penicillin that are not destroyed in the stomach. The finding that penicillin remained intact when it was injected into animals, and that it was not readily destroyed, was an important one because it was a clear indication that it would retain its antibacterial properties in the body.

In May 1940, Florey carried out a crucial experiment, one that Fleming had failed to do twelve years earlier. This experiment was to provide the greatest boost to the chemotherapeutic aspects of the Oxford group's research. He injected eight mice with a lethal dose of bacteria. Four of these mice were placed to one side; the other four mice received injections of penicillin. Florey and Heatley kept a close vigil on these mice: the results of this experiment would provide important information about whether penicillin could cure bacterial infections in living animals. As night approached, Florey and Heatley took turns to go into the laboratory and check on the progress of the animals. Heatley stayed until the early hours of the morning. By half past three in the morning, all four mice that had not been treated with penicillin had died from the bacterial infection, whereas all four mice given penicillin were still alive. The situation remained the same the next day when Florey, Chain and other members of the research team arrived in the laboratory. 'It's a miracle', Florey said.

Experiments were repeated several more times with many more mice and higher doses of bacteria. Penicillin proved to be effective: it was unquestionably able to clear up bacterial infections in mice. The next question was even more important: could penicillin be used in humans? Florey realised that this had to be the next step, but also that it would require much more penicillin:

an adult human being is three thousand times heavier than a mouse. Penicillin in its semi-pure form was difficult to obtain in large amounts in the Sir William Dunn School of Pathology. When the results obtained with infected mice were published, Florey expected there to be a surge of interest from pharmaceutical companies, but he was disappointed: the full significance of penicillin was not noticed by anyone other than Florey and his team.

Florey decided to manufacture large amounts of penicillin himself in his laboratories. Hundreds of culture vessels shaped like bed-pans and designed by Heatley were purchased from a factory so that massive amounts of *Penicillium* mould could be grown to provide enough penicillin for human trials. Florey's department was turned into a penicillin factory, and several young girls were employed to carry out the routine culture and harvesting of the mould; they were called the 'penicillin girls'. Within a few months enough penicillin was available for tests of toxicity in humans.

Initially, Florey and his team examined what happened to penicillin when it was given to volunteer humans. After the first person who was given penicillin by the Oxford group reacted with a high temperature and shivering, there was some concern, but it was soon realised that the reaction was due to some impurity in the penicillin preparation. The well known chemist, Edward Abraham (b. 1913), a member of Florey's team who was involved in determining the molecular structure of penicillin, developed a simple method for removing this impurity. When the contaminant was removed, penicillin was injected into volunteer members of Florey's research group as well as other scientists at the Sir William Dunn School of Pathology. It was found to be harmless to humans, and it was excreted intact in urine and destroyed in the stomach, just as it was in mice and other laboratory animals. These results were promising and Florey was convinced that penicillin could cure bacterial infections in humans as it did in mice.

It was an obvious step now to investigate the effects of penicillin on hospital patients with bacterial infections. A number of such patients were given penicillin injections. The first patient on whom penicillin was tried was a policeman who had extensive bacterial infections on his face; a small infected scratch in the corner of his mouth, originally caused by a rosebush, had spread around his face. He had had an eye removed because it was badly infected, and sulpha drugs were useless against the infection. Several injections of penicillin proved to be beneficial: the infections became less severe and the patient's high temperature fell. However, penicillin was in short supply after several injections and Florey and his colleagues even had to isolate it from the patient's urine and re-inject it into him. The patient eventually died after the infection spread to his lungs. Nevertheless, there clearly had been some improvement and there had been a problem with the supply of penicillin, so there was cause for optimism.

Soon afterwards, when much more semi-pure penicillin was available, a number of other patients were injected with penicillin and this time with great success. A fifteen-year-old boy with a surgical wound infection showed enormous improvement when he was given penicillin; a labourer's back infection cleared up; and a six-month-old baby was cured of a life-threatening infection. Penicillin did, indeed, cure bacterial infections in humans. From pure research, Florey, Chain and their colleagues had rediscovered an antibacterial substance that was to change the world.

Larger trials of penicillin in human patients needed even greater amounts of penicillin and Florey realised that the Sir William Dunn School of Pathology was being stretched to its limits in its capacity as a penicillin factory. He therefore turned to industry and a few companies in Britain did produce reasonable amounts of penicillin for Florey. However, the degree of interest in penicillin by British companies was disappointing, despite the fact that Florey and his team had demonstrated that penicillin could cure human infections and that the Second World War was under way, with bacterial infections being a major problem amongst injured soldiers.

Nevertheless, penicillin did become an important consideration as a 'positive' weapon of war. Indeed, there was some concern that the Germans might produce it in bulk and use it to treat their own injured soldiers. There certainly was some interest in penicillin in Germany. However, the real problem of increasing the yield of penicillin from *Penicillium* cultures had not yet been solved.

In 1941 Florey and Heatley went to the USA to try and elicit some commercial interest in penicillin there. The interest was forthcoming and industrial penicillin production became a reality in the USA. A laboratory belonging to the US Department of Agriculture in Peoria, Illinois, made a significant step forward in increasing yields of penicillin by growing *Penicillium* in the liquid obtained by soaking corn in water. Also, new types of *Penicillium* mould were found, that produced more penicillin than the one discovered by Fleming: one of these moulds, which turned out to be particularly important, was discovered growing on a cantaloupe melon in a Peoria market. The mould from this melon was used as the source of most of the world's commercial penicillin for more than a decade.

Large scale production of penicillin in the USA began at about the same time that the Japanese attacked Pearl Harbor (December 1941). The success of penicillin in treating bacterial infections in civilians and wounded soldiers was such that it gained widespread acceptance. Penicillin was eventually produced world-wide and still is one of the most important antibiotics. Many variants of the original drug have been made. Penicillin is a rather small molecule (Figure 22) consisting of a region called the **penicillin nucleus** and a **side-chain**. The side-chain can be varied tremendously to produce a vast range of different penicillins with diverse properties and effectiveness against different bacteria.

Penicillin shows exquisite discrimination between animal and bacterial cells, which accounts for its low toxicity to humans and its potent antibacterial properties. The reason why penicillin attacks bacteria but not human cells is that it inhibits the ability of bacteria to build their cell walls, structures that cover the bacterial surfaces and protect them from damage. Human and other animal cells lack such cell walls – they have other structures on their surface that protect them – and so penicillin has no effect on them. Penicillin does not directly kill bacteria: it prevents them from making their cell walls and eventually they die as a result of damage and being unable to multiply properly. It used to be

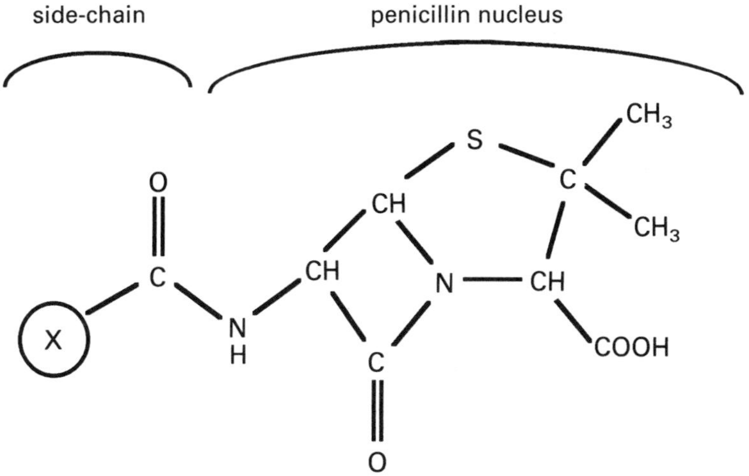

side-chain   penicillin nucleus

Figure 22. Molecular structure of penicillin. The molecule contains a 'nucleus' and a side-chain, X–C=O. By varying X, diverse penicillins with different properties can be obtained.

thought that any drug that killed bacteria would also damage human cells, but penicillin and other antibiotics put an end to this notion. Nowadays it is usual to look for differences between bacteria and humans so that new drugs can be developed that attack structures specific to bacteria. Penicillin, which performs this task almost to perfection, was discovered by chance. All of the scientists involved – Fleming, Florey, Chain, Heatley and their colleagues – were initially spurred by curiosity and not by a deep desire to cure human diseases.

15
DNA, the alphabet of life

F leming's fortuitous discovery of penicillin in 1928 stemmed
from his research on the bacteria that cause boils. At about
the same time, another British bacteriologist, Fred Griffith
(1879–1941), who was studying the bacteria that cause pneumonia, made a discovery that led to one of humankind's greatest
scientific achievements of the twentieth century – the identification of the molecular nature of the genetic material, **deoxyribonucleic acid**, or **DNA**. The significance of Griffith's work was
not at first realised, and, as with penicillin, well over ten years
elapsed before the true importance of Griffith's find was appreciated. It so happened that during the Second World War both
penicillin and the genetic material were revealed to the world.
Whilst penicillin was then accepted as a major step forward, the
significance of the work on the genetic material was still not
widely appreciated and it took many more years before its true
place in history was established.

The discovery of the molecular nature of the genetic material
– the alphabet of life, consisting of sequences of chemical 'letters'
in DNA that code for the proteins and other molecules that
constitute all living organisms – was not initially anticipated by
the scientists involved. (See Chapter 17 for more information
about how DNA encodes proteins.) Nobody could have guessed
that studies of pneumonia-causing bacteria would provide the key
to the chemical composition of the genetic material. Here
is another example of research that gave rise to a completely
unexpected discovery with outstanding and far-reaching significance.

For thousands of years humans have marvelled at the seeming
complexity of themselves and other organisms, and at the physical
resemblance of children to their parents, grandparents, brothers
and sisters. Perhaps understandably, they have frequently invoked
supernatural and religious explanations for these observations.
Now, however, we have the scientific explanation for these geneti-
cally determined features, and it turns out to be more wonderful
than anyone could imagine. A molecule called DNA, which consists of two helices coiled around each other (Figure 23), codes
for the physical features of living organisms. It can be considered
a supreme achievement that the genetic code has been elucidated.
Like the discoveries explaining the origins of the Universe (Chapter 7), the breakthroughs that have led to identification of DNA
as the genetic material reach the heart of human inquisitiveness.
We are nearer than ever to answers to such questions as, 'where
do we come from?' and 'what is life?'

Knowledge of the genetic material has also led to avenues of
research beyond our wildest dreams. DNA can now be manipulated in test tubes and the **recombinant DNA technology** that
has been developed to carry out this work is providing answers
to many biological questions, including those pertaining to the
mechanisms by which a fertilised egg develops into a baby and
the molecular processes involved in human thought. Applications
of recombinant DNA technology abound. Examples include: the
development of DNA fingerprinting (Chapter 17); studies of the
processes by which normal cells become cancerous; increased
understanding of inherited diseases such as cystic fibrosis and

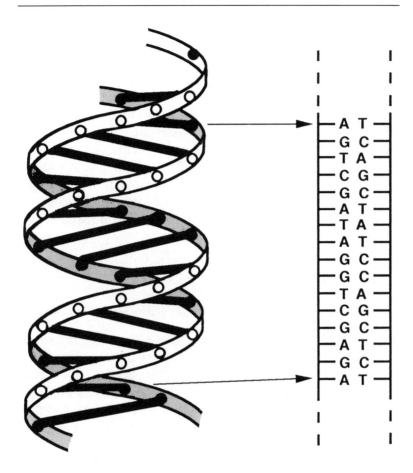

Figure 23. DNA (deoxyribonucleic acid) consists of two helices wound around each other (left). Sugar and phosphate form the curves of the helices; bases form the 'rungs' between them. There are four types of base in DNA: adenine (A), guanine (G), cytosine (C) and thymine (T); A always pairs with T and G with C (right). The sequence of bases determines how the genetic code is 'read' by cells into proteins.

muscular dystrophy; and production of new drugs for treatment of a wide range of diseases. Recombinant DNA technology is here to stay and future generations will benefit enormously from its fruits.

In view of the rapidity with which our understanding of DNA and the mechanisms by which it codes for the physical features of living organisms has progressed, it is hard to believe that scientists thought that DNA could not possibly be the genetic material well into the twentieth century. Indeed, even in 1950 there were many respected biologists who thought that DNA was a relatively unimportant substance and that genes must be composed of proteins. DNA was a simple chemical, whereas proteins were much more complex and were already known to exhibit a diversity that could easily explain their ability to code for the many features of a living organism. The road to the acceptance of DNA as the genetic alphabet begins in 1869, with the discovery of DNA.

Genes, DNA and chromosomes

The Swiss scientist, Friedrich Miescher (1844–1895), discovered DNA in 1869 when he chemically treated human white blood cells, which he obtained from the pus of septic surgical wounds. The cells became viscous and sticky, and Miescher showed that the viscous substance was present in the nucleus of the cell, rather than in the cytoplasm that surrounds the nucleus. He called the substance 'nuclein' (it later became known as DNA) and showed that it contained phosphorus and was acidic in nature. When it later became clear that the nucleus contains the cell's genetic material, most scientists believed that proteins associated with the DNA, rather than the DNA itself, carried the genetic information.

During the nineteenth century most biologists thought that genetic characteristics of an individual were inherited from its parents in blood or other fluids that were passed from parents to offspring. This explains the modern use of terms such as 'blue blood' and 'blood relative'. However, the Austrian monk, Johann Gregor Mendel (1822–1884), carried out some elegant experiments on pea plants and the results of his work proved the blood theory wrong. Mendel showed that the factors (genes) that carry information for inherited physical traits, such as pea shape, flower colour or stem height, do not blend with each other. For example, a tall pea plant crossed with a short one does not produce inter-mediate sized plants; rather, it produces either tall or short off-spring. Likewise, the eyes of a child whose mother and father have blue eyes and brown eyes, respectively, are not a blend of the two colours. Mendel also showed that each parent transmits half of its genes to each of its offspring, so that the offspring contain essentially equal numbers of genes from each parent. However, each offspring receives a different combination of its parents' genes from that received by his or her brothers or sisters. At the time, Mendel had no idea what genes were made of or where in an organism's cells they were: he simply had evidence for their existence and properties from his studies of their visible expression.

Unfortunately, Mendel's pioneering results were neglected until 1900, when they were rediscovered and confirmed. The nuclei of living cells were subsequently found to contain pairs of structures called **chromosomes**, which could be seen under a microscope and which doubled in number just before a cell divided. After doubling, half of the chromosomes were trans-mitted to each of the two 'daughter' cells, which therefore con-tained the same number of chromosomes as the parent cell – exactly what would be expected of genes. It was subsequently confirmed that chromosomes contain the genes, and sperm and egg cells were found to contain half of the number of chromo-somes found in other cells of the body: during fertilisation the chromosomes from the sperm and those from the egg come together into the single nucleus of the fertilised egg. Chromo-somes contain protein as well as DNA and it was generally accepted that the proteins were the carriers of genes.

The chemical composition of DNA was known by the turn of the twentieth century. DNA is composed of three basic chemical constituents: a sugar called deoxyribose, a phosphate, and a chemical group called a base. The sugar and phosphate are the same throughout the DNA molecule, whereas the base can be one of four types. At the beginning of the century nobody knew what the molecular structure of DNA looked like: they knew only about its composition. However, the constituents appeared to be simple and it was very difficult to imagine how DNA could possibly code for the complexities of a living organism.

When the molecular structure of DNA was elucidated in the early 1950s by Francis Crick (b. 1916), James D. Watson (b. 1928), Maurice Wilkins (b. 1916) and Rosalind Franklin (1920–1958), its double helical structure immediately offered an explanation for the mechanism by which genes are passed on to daughter cells when a cell divides; and it provided clear indications as to how such an apparently simple molecule can encode enormous amounts of information. The two helical strands could unwind and new, complementary strands could be made using each of the unwound strands as a template, so that the copying is accurate. Following replication of each parent strand, the two new pairs of strands could be transferred, one to each of the two daughter cells. Four possible bases arranged in a long strand containing a sequence of thousands of such bases or 'letters' could easily provide the necessary 'sentences' to code for the complexity required of living organisms.

Nevertheless, progress in determining the molecular nature of the gene was extremely slow during the first four decades of the twentieth century, and when the crucial discoveries that led to its identification did arise, they did not come from scientists who were deliberately searching for the genetic material. Rather, the chemical nature of the gene was identified as a result of the work carried out by Fred Griffith in Britain and by a group of US scientists, all of whom were interested in the bacteria that cause pneumonia. It was not until very late in these studies that the scientists involved realised they had, in fact, identified the alphabet of life.

Pneumococci, Griffith and transformation

At the beginning of the twentieth century pneumonia was one of the major health problems in developed countries, rivalling cancer and heart disease today. Most cases of pneumonia were caused by bacteria called pneumococci, and many research groups were involved in studying these bacteria in the laboratory using the methods of bacteriology that were pioneered by Pasteur, Koch and their contemporaries.

It was known that pneumococci existed in various forms, which could be distinguished by their ability to be recognised by anti-bodies (Chapter 18) specific to each type. Thus, antibodies that recognise one type of pneumococcus would not recognise any of the other types of pneumococcus. Another form of variability occurred in which pneumococci could be converted to forms that were unable to produce pneumonia in mice. This change could be detected easily by growing the bacteria in a dish on the surface of a jelly containing nutrients: each pneumococcus bacterium produced a visible colony, containing millions of pneumococci, on the surface of the jelly. The infectious forms of pneumococci grew as large, smooth colonies that glistened, whereas the forms that failed to produce pneumonia infections grew as smaller, grainy colonies with a rough appearance. On the basis of their colony appearance, the infectious forms of pneumococci were called smooth, or S pneumococci, whereas the non-infectious forms were called rough, or R forms.

The fact that smooth (infectious) types of pneumococci could be converted to rough (non-infectious) types was considered by many scientists to be of great importance in understanding the mechanisms by which these bacteria cause disease. If the bio-chemical difference between the rough and smooth forms could be understood, molecules involved in production of infectious pneumonia might be identified. Perhaps inhibition of the syn-thesis of these molecules with a drug would be a means of treating pneumonia. Indeed, it was discovered that the smooth forms of pneumococcus have a thick outer coat on their surface, whereas the rough forms lack this **capsule**. The pneumococcal coat was eventually shown to consist of a substance called a **polysacchar-ide**, which is a large molecule made up of sugar units. The scientific term for a sugar unit is a **monosaccharide**; polysac-charides contain many hundreds or thousands of such sugar units joined together in a network of long chains. Polysaccharides fre-quently perform structural and protective roles in cells; they include cellulose, the constituent of plant cell walls, as well as the polysaccharide coating of bacterial cell walls.

The US scientist, Oswald Avery (1877–1955), who worked at the Rockefeller Institute for Medical Research in New York, was a leading researcher of pneumococci. His team demonstrated that the surface capsule of pneumococci consists of polysaccharides. Avery called pneumococcus the 'sugar-coated microbe'. He was especially interested in the chemistry of the capsule and in its possible therapeutic applications to pneumonia.

Fred Griffith, who worked with the British Ministry of Health in London, began studying pneumococci after the First World War. He isolated pneumococci from the sputum of pneumonia patients and grew them in his laboratory in culture flasks or dishes containing appropriate nutrients; he also examined their ability to produce pneumonia infections in mice.

Griffith noted that sputum from a single pneumonia patient often contained four or five different types of pneumococci. He had only two possible explanations for this observation: either these patients had been infected on four or five different occasions, each time by a different type of pneumococcus, which was unlikely, or this was another example of bacteria within a population of pneumococci being able to change themselves somehow to generate several different types.

As the rough (R) forms of pneumococci lack the polysaccharide coating, and the smooth (S) forms have a thick capsule, it seemed that this coating might be crucial in endowing pneumococci with infectivity. Griffith decided to study the mechanism by which S forms of pneumococci were converted to R forms.

It became clear to Griffith that the non-infectious R forms of pneumococci could readily be obtained in the laboratory from infectious S forms using various procedures. What was also clear, in agreement with previous experiments carried out by other scientists, was that all R forms lacked the polysaccharide capsule, whereas all S forms had a capsule. Whilst S forms always produced pneumonia infections when they were injected into mice, R forms usually were not infectious.

However, Griffith did find that when he injected very large numbers of R forms into mice he could almost always produce pneumonia infections. Lower levels of most R pneumococci failed to infect mice, even though the same low number of S forms always produced the disease. Griffith provided a possible explanation for this phenomenon. It could be, he said, that R forms contain a small amount of a substance that is normally present in large amounts in S forms and that allows synthesis of the polysaccharide capsule. If large enough numbers of R forms are injected into mice, there might be a sufficient amount of this substance to stimulate capsular formation and therefore allow the R forms to turn into S forms and so produce infection. He had a simple way of testing this idea. If something present in S pneumococci causes them to produce capsular polysaccharide and if this substance stimulates R forms to change to S forms, then heat-killed S forms (which are not infectious) might still contain this substance and cause R forms to become S forms.

After several attempts, Griffith carried out an experiment that demonstrated that a substance in S pneumococci could indeed transform R forms into S forms (Figure 24). He injected eight mice with heat-killed S forms mixed with a small number of R forms. Two of these mice died of pneumonia infection and S pneumococci were recovered from them. In control experiments, neither the heat-killed S bacteria alone nor the R forms alone produced infections at the doses used, showing that the S forms were not surviving the heat-killing process. Although transformation from R to S forms occurred only in bacteria in two of the eight mice, the result was significant when compared to controls, and Griffith managed to reproduce these results in subsequent experiments. It was clear that something present in S bacteria was allowing R forms to become S forms.

When Griffith's meticulous work was published in 1928 it was met with great scepticism, but it was not long before others working on pneumococci confirmed that R pneumococci can indeed be transformed to S forms using dead S forms. Oswald Avery and his colleagues were equally incredulous of Griffith's results, but they also confirmed them. Avery believed that it should be possible to obtain transformation without the use of mice. Griffith had always injected a mixture of dead S and live R pneumococci directly into mice. It should be possible, Avery thought, to mix

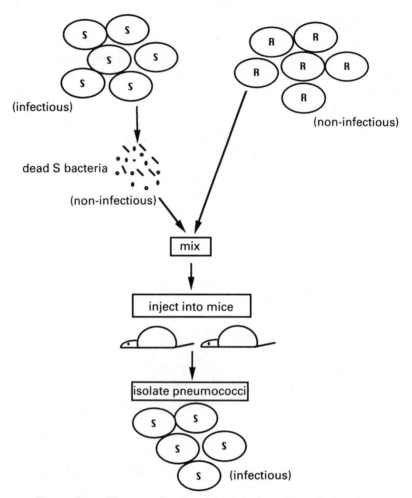

Figure 24.   The transforming principle, later identified as the
genetic material, DNA, was discovered in 1928 by Griffith
whilst he was studying the bacteria that cause pneumonia.
When living, non-infectious bacteria (rough, or R forms) were
mixed with dead, infectious bacteria (smooth, or S forms), and
the mixture was injected into mice, pneumonia occurred and
infectious S bacteria were recovered from the mice. Evidently,
DNA was being transferred from the dead S bacteria to the
living R bacteria and becoming part of the genetic material of
the R forms, endowing them with infectivity.

the two in a test-tube then streak them out on nutrient jelly in a dish: any R forms that had become S forms would appear as smooth colonies on the surface of the jelly. Griffith had tried this but failed. Avery and his collaborators did eventually manage, in 1931, to produce R to S transformation in a test-tube. This set the scene for a much more controlled and detailed study of the mechanism of pneumococcal transformation. The unknown factor present in S pneumococci that allowed R forms to become S forms became known as the **transforming principle**. Avery made it a major aim of his life's work to discover the chemical nature of the transforming principle. Little did he know at the time he embarked upon this research programme that the transforming principle was, in fact, the genetic material used universally by living organisms as the alphabet that encodes the necessary substances and processes that make an organism what it is.

The nature of the transforming principle

The transforming principle was enigmatic. It was not simply a substance present in S forms of pneumococci that transiently converted R to S forms, but it was also inherited indefinitely by bacterial offspring of the transformed bacteria and their subsequent generations. Once an R pneumococcus had been transformed to an S pneumococcus, its offspring and their future generations were also S forms. Heritability was a property of genes. However, most scientists believed that the transforming principle was actually a substance that perhaps switched on existing genes or other latent activities in R forms that were normally only active in S forms. In other words, nobody seriously believed that the transforming principle was the stuff of genes itself. The biochemical nature of genes was completely obscure at the time, and it is not surprising that nobody really appreciated the possibility that the transforming principle might be genetic material that is transferred from S to R pneumococci and subsequently becomes a heritable gene of the transformed R bacteria. It seemed much more likely that heat-killed S bacteria were influencing the R forms without actually transferring a heritable substance directly to the R bacteria.

Avery's team showed once and for all that the heat-killed S forms of pneumococci were truly dead and that the transforming principle was a component of dead S bacteria. He extracted S pneumococci in a way that destroyed the integrity of the bacterial cells and filtered the resulting extract through a sieve that prevented bacteria from passing through it. This extract, which was more pure than previous preparations containing transforming principle, definitely contained no intact S bacteria, yet it still was able to transform R pneumococci to S forms. By 1933, Avery's laboratory had increased the efficiency of transformation of R forms in test-tubes and could treat extracts of S bacteria in such a way that many more impurities were removed, leaving the transforming principle in a purer form. However, the preparations containing the transforming principle were still highly impure and identifying which of the many substances present was the transforming principle was still a long way off.

In 1934, another scientist, Colin Munro MacLeod (1909–1972), joined Avery's team and began to work on the identification of the transforming principle. MacLeod became highly competent at carrying out pneumococcal transformation experiments and he made a major step forward by isolating a new R form of pneumococci that showed an enormously increased frequency of transformation when treated with S extracts containing the transforming principle. This new R strain was used in all subsequent transformation studies and it greatly facilitated the eventual identification of the transforming principle. MacLeod and Avery published their work on transformation in 1935. Although they had failed to identify what substance the transforming principle was, they still believed that it was something that influenced latent activities already present in R forms, rather than that it was a factor which could be inherited that had been transferred from S to R bacteria.

MacLeod and Avery made slow progress over the next two years, although they did find that treatment of preparations containing transforming principles with substances that specifically destroy proteins failed to inactivate the principle. This was the first hint that the transforming principle might not be a protein. Three more years elapsed with little, if any, work being done on the transforming principle. It was not that Avery lost interest; he still believed the transforming principle was an important subject of research. Part of the reason for the lapse of research was that the antibiotic sulpha drugs appeared in 1937 (Chapter 14) and MacLeod and others in Avery's laboratory directed their efforts towards the study of the effects of these drugs on pneumococci.

In 1940 MacLeod and Avery decided that they would return to their studies of the transforming principle and make a renewed effort at identifying its chemical nature. It was clear that very large volumes – tens of litres – of pneumococci would be needed in attempts to purify the transforming principle, and this was, at the time, not an easy task, especially as the S forms were a hazard, putting the experimenter at some risk of contracting pneumonia. New methods and apparatuses were devised to minimise this health risk and eventually large quantities of pneumococci were being produced on a regular basis in Avery's laboratory. MacLeod and Avery discovered that they could obtain the transforming principle in the form of a solid containing essentially no protein, supporting the idea that the principle was not a protein. This was the purest preparation of the transforming principle they had so far obtained: although it lacked any detectable protein, it did contain most of the bacterium's **nucleic acid** as well as most of the polysaccharide (including the pneumococcal surface capsule).

Nucleic acid consists of two types of chemical, DNA (deoxyribonucleic acid) and RNA (ribonucleic acid). At the time it was thought that DNA was the nucleic acid of animals and that RNA was the nucleic acid of plants; pneumococci were thought to contain RNA but not DNA. We know today that all cells have both DNA, which is the genetic material, and RNA, which is involved in the processes by which genes are 'read' by a cell to produce proteins that are necessary for the proper structure and function of the cell. When Avery and MacLeod examined their preparation containing the transforming principle for DNA and RNA, they found that it contained both. This was the first time that DNA had been detected in pneumococci. Could the transforming principle be made of DNA or RNA, which at the time were considered to be of minor biological importance? A substance that specifically destroys RNA but leaves DNA intact was shown not to inactivate the transforming principle in these preparations. If the transforming principle was a nucleic acid it would appear more likely to be DNA than RNA.

The possibility also remained that the transforming principle was identical with the capsular polysaccharide, which was a major component in the preparations containing the transforming principle. Avery and MacLeod had originally thought this was not the case; they preferred to think that the transforming principle was some other substance that simply stimulated enzymes to synthesise new capsular polysaccharide in the newly transformed R bacteria. However, it was still possible that the polysaccharide obtained from S forms could itself stimulate enzymes in R forms to produce new capsular polysaccharide.

In 1941, MacLeod left Avery's laboratory to take up a post elsewhere and Avery was fortunate to have a new member of his laboratory, Maclyn McCarty (b. 1911), join him at the same time. McCarty carried on where MacLeod had left off, although MacLeod occasionally came back to the laboratory to assist with the research. It was clear at this time that Avery's team were getting nearer to the identity of the transforming principle. It is still unlikely that at this stage the scientists involved realised the full significance of their work: several more years elapsed before any of them seriously considered the possibility that they had identified the genetic material.

McCarty initially was concerned whether capsular polysaccharide, which conferred infectivity on S pneumococci, might also be the transforming principle. He eliminated this possibility by showing that substances that specifically degrade polysaccharide do not inactivate the transforming principle. In addition, the amount of polysaccharide present in preparations of transforming principle could be reduced by growing the S bacteria in certain culture broths: this did not reduce the yield of transforming principle.

In 1942 McCarty showed that partially pure preparations containing transforming principle were not affected in their capacity to transform R pneumococci to S forms after the preparations had been treated with substances that specifically destroy polysaccharides. With proteins, RNA and polysaccharides eliminated, there was one main candidate left for the transforming principle: DNA.

It must have been about this time that Avery and his colleagues seriously began to wonder whether or not the transforming principle was indeed DNA. Since DNA was also known to be associated with chromosomes, which carried the genes of a cell, and since the transforming principle behaved in a heritable manner, it is likely that the idea was emerging amongst members of Avery's team that they had discovered the chemical nature of the gene. Certainly Avery was beginning to think in this way. In 1943 he wrote, in a letter to his brother, 'If we are right, and of course that's not yet proven, then it means that nucleic acids are not merely structurally important but functionally active substances in determining the biochemical activities and characteristics of cells – and that by means of a known chemical substance it is possible to produce *predictable* and *hereditary* changes in cells. This is something that has long been the dream of geneticists.' In the same letter, he clearly states that the transforming principle 'sounds like a virus – may be a gene' and he goes on to say 'it takes a lot of well documented evidence to convince anyone that [DNA] could possibly be endowed with such biologically active and specific properties.' He had the task ahead of him of convincing other scientists, and perhaps even himself, that not only had his group identified the substance of genes, but also that it was DNA rather than a protein.

At about the same time, methods were being developed for the isolation and purification of DNA: some scientists were interested in characterising its properties and functions, although none of them were motivated by the possibility that genes were made of DNA. Avery and McCarty discovered that these methods were also useful for preparing the transforming principle, further supporting the idea that the principle was DNA. The two scientists became involved in looking for evidence to support their contention that the transforming principle was DNA and that it was not any of the other well known biochemical substances such as proteins or polysaccharides. They eventually obtained a preparation of the transforming principle that appeared to be completely free of any substance other than DNA. Its chemical composition was confirmed to be that of DNA; its ability to transform R pneumococci to S forms was enormous; and no matter how hard they looked they could not find any minor impurities that might be the transforming principle. A substance was also available that specifically destroys DNA but not RNA, proteins, polysaccharides or other chemicals. This substance inactivated the transforming principle and provided further strong support for its DNA nature. The work was published in 1944.

What seemed to be happening in transformation of R forms of pneumococci to S forms is that the R forms lack a gene or genes necessary for production of the bacterium's sugar coat. When the DNA with the region that encodes this gene is extracted from S forms and added to R forms, the R forms take up the gene and incorporate it into their own genetic material: it becomes a part of the R forms. These R forms are then able to make the sugar coat and this makes them infectious S forms.

Many scientists at the time failed to appreciate the significance of Avery, MacLeod and McCarty's publication. The biochemical nature of the genetic code was staring everyone in the face, but few even noticed or believed it! However, those who did see the significance of the work did much to further the idea that DNA is the genetic material. Some scientists changed the direction of their own research in order to concentrate on DNA because they believed it was probably the genetic material. Sperm and egg cells were found to contain half of the amount of DNA that was present in other cells of the body, which was consistent with the idea that half the number of an offspring's genes came from each parent. Precise chemical studies of DNA from various organisms were carried out and these provided a better knowledge of the chemistry of DNA and how it varies from one organism to another. It became clear that DNA was not as simple as was previously thought, and scientific arguments were put forward to support the idea that it could, after all, theoretically code for the many complex components of living organisms.

Other scientists eventually obtained the evidence necessary to corroborate Avery, MacLeod and McCarty's work and strengthen the theory that DNA was indeed the genetic material. One particular experiment that argued the case strongly was carried out by Alfred Hershey (b. 1908) and Martha Chase, who, in 1952, showed that the genetic material of bacterial viruses was DNA. These viruses, or **bacteriophage**, which specifically infect bacteria, contain protein and DNA. It was known that part of the bacteriophage enters the bacteria that they infect whilst another part (the bacteriophage coat) remains outside of the bacteria. The genetic material of the bacteriophage takes command of the bacterial cell, directing it to produce more bacteriophage: in this way the bacteriophage multiplies within the bacterium. It is the genetic material of the bacteriophage which enters the bacterium: the virus coat is involved only in binding the bacteriophage to the bacterium. Because DNA and protein are chemically distinct, they can be differentially tagged so that they can be distinguished from each other. Hershey and Chase labelled bacteriophage by incorporating radioactive atoms that specifically tag either DNA or protein. This allowed them to follow, separately, the DNA and protein of bacteriophage during and after infection of bacteria. They found that the DNA entered the bacterial cell whereas the protein remained outside in the bacteriophage coat (Figure 25). This meant that DNA, but not protein, encodes the information needed to direct the making of new viruses within the bacterium. In other words, DNA is the genetic material of bacteriophage.

In 1953, the year after Hershey and Chase reported their experiment, Crick and Watson discovered and published their famous double helical structure of DNA. Everything fitted: DNA clearly was the genetic alphabet of life, and its molecular structure revealed how beautifully and elegantly a substance as chemically simple as DNA can carry the millions of pieces of information necessary to describe a living organism. Once DNA had been established as the genetic material and its molecular structure had been elucidated, a new era of biology began. Nowadays DNA can be manipulated at will in a test-tube, its alphabet can be read using chemical methods that determine the order of 'letters' along a length of DNA; and the proteins it encodes can be found by artificially putting it inside living cells. Recombinant DNA

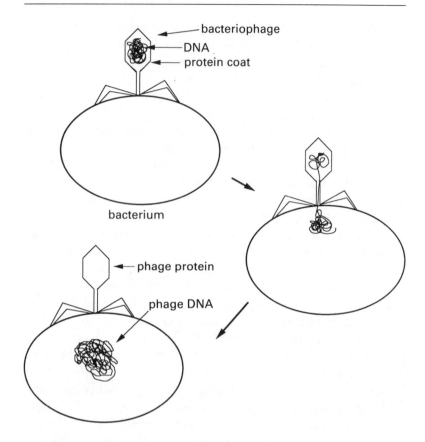

Figure 25.   Hershey and Chase confirmed the idea that DNA is the genetic material in the early 1950s. They 'labelled' the DNA of one bacteriophage population and the protein of another population and allowed them to infect bacteria. Only the labelled DNA entered the bacteria: the labelled protein remained outside the bacterium. This demonstrated that bacteriophage DNA is the part that enters bacteria and allows more copies of the phage proteins and DNA to be produced. In other words, all of the information for phage production – the phage genetic material – is present in the injected DNA.

technology is here to stay, with its promise of solutions to fundamental questions concerning the nature of life and its immense potential applications to medicine.

The 1944 publication by Avery, MacLeod and McCarty is considered to be one of the great classics of twentieth century biology. It presents the first clear evidence that DNA is the true stuff that genes are made of. Yet, none of these scientists received a Nobel Prize for their work. The main reason for this is probably that nobody appreciated just how important the discovery was: there was too much opposition to the idea that DNA was the genetic material. Time has shown that Avery and his colleagues made one of the most important discoveries in the history of biology. In a sense, it does not matter how their work was received at the time it was published: Avery, MacLeod and McCarty are now considered to be the discoverers of the chemical nature of the alphabet of life. Their work is another clear example of a serendipitous breakthrough that provided a key to the answer to one of the most important scientific questions that humans have ever asked, 'what is life?'

16
Cutting DNA with molecular scissors

W hen the transforming principle was finally demonstrated to be DNA (Chapter 15), it not only revealed the chemical nature of the genetic material, but also indicated that DNA could be taken up by bacterial cells from their surroundings and become part of their genetic material. Indeed, the transforming principle was DNA encoding the genetic information required for synthesis of the sugar coat of pneumococcal bacteria. This DNA, when isolated from infectious pneumococci (S forms), could be taken up by live, non-infectious pneumococci (R forms). The DNA would then be incorporated into the genetic material of the R forms, which would consequently be heritably transformed into infectious S forms.

Foreign DNA can also enter bacterial cells if it is packaged into bacterial viruses. Just as viruses invade the cells of humans and other animals they can also invade bacterial cells – the ones that do so are called **bacteriophage**. Bacteriophage were discovered by the British scientist, Frederick Twort (1877–1950) in 1915, and the French-Canadian bacteriologist, Felix d'Herelle (1873–1949), in 1917. These scientists noticed that some of their bacterial cultures contained an infectious agent that destroyed the bacteria. This infectious agent was smaller than known bacteria since it could pass through filters that trapped bacteria, and it was similar in this respect to viruses that had been shown to cause animal and plant diseases. D'Herelle coined the name, 'bacteriophage', which means 'bacterium-eater'.

It later became clear that bacteriophage, also known simply as **phage**, were indeed viruses that infect bacteria. Phage were shown to have their own genes, encoded by their DNA, which are carried in a container made of proteins that form the phage's 'head'. A phage attaches itself to a bacterium by its 'tail', which is also made up of proteins, and injects its DNA into the bacterium. The protein head remains outside the bacterium and the DNA, which carries information required for the phage to multiply and produce many more phage offspring, enters the bacterium and directs the production of more phage (Figure 25, Chapter 15). The genes that constitute phage DNA encode proteins that 'deceive' the bacterium's cellular machinery to make more phage: in this way, the phage takes over control of the bacterium in order to reproduce itself. The phage DNA also encodes structural proteins that make up the head and tail of the phage: these genes are 'read' to produce the structural proteins inside the bacterial cell.

Phage infect only bacteria: they are harmless to cells of other organisms. At one time it was thought that phage might be potential antibacterial agents, but this proved not to be the case. However, they became very useful for investigating bacterial biochemistry and genetics; they also opened up the field of phage genetics, which contributed tremendously to our knowledge of genetics in general. More recently, phage have become powerful tools in recombinant DNA technology, for isolating and manipulating genes from humans, other animals and plants. Recombinant

DNA technology has revolutionised biology and medicine. Almost any gene can be isolated from any chosen organism and its sequence of chemical 'letters' determined, offering detailed knowledge about the DNA encyclopaedia of the organism. This may provide diagnostic reagents and possible drugs and vaccines on an unprecedented scale. Bacteriophage have played an important role in this revolution. By incorporating genes encoding proteins of humans or other organisms into the phage's own DNA, scientists can trick the phage into accepting the foreign genes as if they were its own. When these recombinant phage then invade bacteria, they multiply, along with their foreign DNA. In this way, phage can be exploited in order to isolate and produce DNA encoding genes of other organisms. The foreign genes can be chosen to make specific proteins (encoded by the foreign DNA) when the phage invade bacterial cells. In this situation, recombinant phage may be a source of production of large amounts of proteins. Vaccines, diagnostic proteins and other medically useful proteins, such as hormones, can be produced in bacteria by this method.

Whilst bacteriophage are tools in their own right for handling and investigating genes and their protein products, studies of their basic biological properties also led to the discovery of another group of substances, the **restriction enzymes**, which have contributed crucially to our ability to manipulate genes. Enzymes, which are usually proteins, speed up the rates of chemical reactions without themselves being altered in the process – they are the catalysts of living organisms. Restriction enzymes are proteins that cut DNA molecules at precise sites to produce clearly defined fragments. It is this precision that allows chosen genes to be cut away from other genes so that they can be isolated. Restriction enzymes allow stretches of DNA to be mapped: fragments of DNA produced by restriction enzyme digestion can be separated from each other and their positions with respect to each other can be determined. Without restriction enzymes there would be no recombinant DNA technology. They are the molecular scissors that allow segments of DNA to be excised from the DNA encyclopaedias of living organisms, and that allow two or more pieces of DNA to be snipped and joined together in precise ways. They have become a routinely used tool of molecular biologists, and are as much a part of recombinant DNA research as a saw is to carpentry.

Restriction enzymes occur naturally: they are one of several groups of Nature's enzymes that biochemists have isolated and exploited as tools. The accuracy, precision and versatility with which restriction enzymes cut DNA into clearly defined pieces cannot be equalled by any machine or chemical reaction invented by humans. Yet when restriction enzymes were discovered, few scientists realised how important they would become. Indeed, many people thought that research on these enzymes was highly esoteric and was unlikely to be of any practical value. As with many other discoveries, however, restriction enzymes proved the point that pure research can and does produce unexpected applications of great value. Many scientists were involved in the identification, characterisation, and exploitation of restriction enzymes. Three of them – Werner Arber (b. 1929), Daniel Nathans (b. 1928) and Hamilton O. Smith (b. 1931) – received the 1978 Nobel Prize in Physiology or Medicine for their contributions.

Discovery of restriction enzymes

Restriction enzymes were discovered as a result of several intriguing experimental findings regarding the infection of bacteria by phage. In the early 1950s a number of different groups of scientists noticed that some strains of bacteria were not as prone to infection by particular phage as other strains of bacteria of the same species. In other words, if phage were added to one strain of bacterium, they would multiply prolifically within the bacteria and produce millions of phage offspring, but if the same phage were added to bacteria of a different strain they would hardly multiply at all. Because the phage were the same in both cases, the idea arose that this phenomenon occurred as a result of differences between the bacterial strains. Strains of bacteria that prevented the phage from multiplying within them were said to be **restrictive** for phage infectivity.

Phage that were able to grow well in one particular (non-restrictive) bacterial strain were nevertheless able to grow to some degree, albeit poorly, in restrictive strains of bacteria and small numbers of phage could be recovered from these restrictive bacteria. When these phage were added to the same restrictive bacteria, however, they were now able to multiply rapidly: they were no longer restricted by the bacteria (Figure 26). Similarly, if the small number of phage recovered from restrictive bacteria were added to the strain of bacterium in which the phage originally multiplied well, they now grew poorly. In other words, phage could be made to grow well in a restrictive bacterial strain if they were passed through an infection cycle in these bacteria. One interpretation of this data was that bacteria modify phage in some way and that this modification allows the phage to grow in the bacteria.

Studies of these **restriction–modification systems** in different bacteria showed that they were common – many bacteria possessed them. They were also very often specific to each bacterial strain; when a phage was modified by one bacterial strain it could then grow specifically in that strain, not in any other strains. It also became clear that some bacterial strains had more than one restriction–modification system and so could modify phage in more than one way.

The mechanisms by which phage were modified to allow them to multiply in bacteria, and restricted to prevent them from multiplying, were not understood until the 1960s, when the Swiss biochemist, Werner Arber, and his colleagues, showed that it was the phage DNA, rather than the proteins of its head or tail, that was modified by bacteria. Arber and his team noticed that the DNA of phage grown in a non-restrictive bacterial strain was degraded soon after it entered the restrictive bacteria. The same DNA was not degraded if it entered non-restrictive bacteria. Phage were able to attach to restrictive bacteria by their tails and inject their DNA into those bacteria, but the phage DNA did not survive for very long inside the restrictive bacteria. This concentrated effort on studying the phage DNA, how it was modified, and why a failure to be modified meant that the DNA would be degraded by restrictive bacterial strains.

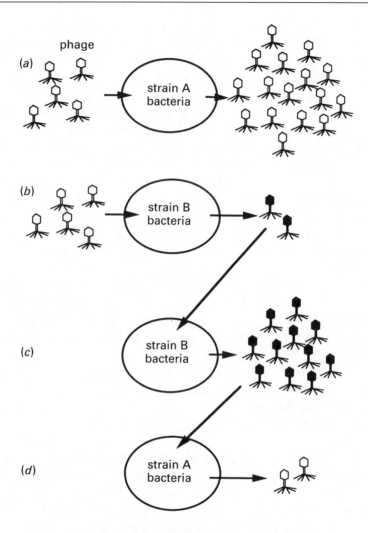

Figure 26.   Bacterial restriction of phage infections. Phage
can multiply well in bacteria of strain A (*a*), but not well in
bacteria of a restrictive strain, B (*b*). If the small numbers of
phage produced in strain B are now used to infect strain B
bacteria, restriction no longer occurs (*c*): strain B has somehow
modified the phage to allow it to multiply. The phage that have
passed through strain B bacteria are now restricted by strain
A (*d*).

The fact that phage DNA was degraded when it entered restrictive bacteria suggested that proteins called **DNA endonucleases** were involved in bacterial restriction of phage multiplication. DNA endonucleases are enzymes that degrade DNA; several similar enzymes were already known and they degraded DNA extensively into small fragments. However, detailed studies of degradation of phage DNA by restrictive bacteria indicated that the phage DNA was initially chopped into relatively large fragments, and these large fragments were then broken down into smaller pieces. The initial degradation of phage DNA into large fragments appeared to be precise: the evidence available suggested that the DNA was initially chopped into defined pieces rather than being broken down into random-sized fragments.

Studies of the mechanisms by which phage DNA is modified by bacteria led to the discovery of specific enzymes that add chemical groups to DNA: these enzymes are called **DNA methylases**. Bacteria can degrade phage DNA using restriction enzymes (restriction DNA endonucleases) in order to prevent an infection with phage. To protect themselves from their own restriction enzymes bacteria have to modify their own DNA, which is done using DNA methylases. If a phage does survive restriction, it becomes modified by the bacterium's DNA methylases in the same way that the bacterium modifies its own DNA. The modified phage DNA can now replicate in the bacterium, since it is no longer susceptible to the restriction endonucleases. This explains why passage of phage through restrictive bacteria produces phage offspring that will now multiply well in these previously restrictive bacteria: the phage DNA has been modified to protect it from restriction enzymes and as it is no longer degraded, the phage can reproduce successfully.

It was later demonstrated that restriction and modification of phage DNA could be detected using extracts of bacteria instead of live bacterial cells. This made the route to elucidating the mechanism of restriction–modification easier. The bacterial extracts contained active DNA methylases that modified phage DNA, and they contained restriction enzymes that degraded unmodified phage DNA. The DNA methylases and restriction enzymes could now be purified from bacterial extracts in isolation from all other bacterial components.

The US biochemist, Hamilton O. Smith, and his co-workers, were the first to isolate a restriction enzyme and to show that it cleaved phage DNA into defined fragments. Daniel Nathans, who worked in the same laboratory, was studying a cancer-causing virus and he wanted to isolate the DNA fragment that was responsible for its cancer-causing properties. He obtained some of Smith's enzyme and successfully isolated this fragment of DNA from the virus. Nathans and his team treated phage DNA with the restriction enzyme to produce eleven DNA fragments, and showed that each of these fragments contained one or more genes encoded by the phage. Nathans produced the first **restriction map**. Rather than chopping the phage DNA just anywhere along its length, the restriction enzyme was cutting it at very precise sites.

Similar precision cutting was shown to be a feature of many more restriction enzymes, and it meant that these enzymes must be recognising very specific regions along the length of the DNA: they must be cutting the DNA within defined sequences of the four-letter genetic code (the letters, or bases, that make up the genetic codes are A, T, C and G). When methods identifying the bases along a DNA molecule were developed, the base sequences recognised and cut by restriction enzymes were determined. It was found that there were many possible sequences that could be cut by the enzymes, and that each restriction enzyme snipped DNA at its own characteristic base sequence. However, different restriction enzymes sometimes recognised the same sequence. The availability of restriction enzymes that chop DNA at exact points along its length provided a means of cutting genes out of the DNA encyclopaedias of living organisms and of cutting genes into smaller pieces. Concomitant with other important developments in methodology, such as techniques for isolating DNA according to their size and for detecting specific DNA sequences in a background of other DNA sequences, restriction enzyme cutting of DNA molecules gave scientists a new approach to handling and studying genes in the laboratory. By 1990, well over a thousand different restriction enzymes, recognising hundreds of different DNA sequences, had been characterised, and several hundred were available commercially.

Properties of restriction enzymes

Restriction enzymes are named after the organisms from which they are isolated. The first two letters of the species component of the Latin name of the organism are added to the first letter of the genus name. For example, the bacterium, *Escherichia coli*, is a member of the genus *Escherichia*, and has the species name *coli*. Restriction enzymes obtained from *Escherichia coli* will therefore be called Eco. The enzyme, EcoRI, for instance, was the first restriction enzyme that was isolated from strain R of *Escherichia coli*. Most restriction enzymes have been found in microorganisms, but they may be much more generally distributed than that. Indeed, one has been detected in human embryos: humans belong to the species, *Homo sapiens*, and the human restriction enzyme is called HsaI. Whether all restriction enzymes have a role in protecting cells from foreign DNA, such as viral DNA, remains to be determined: it is possible that some of them might have precise functions in controlling the structure, organisation, properties and decoding of genes.

Some restriction enzymes do not cut DNA at precise sites, and these are not particularly useful tools for recombinant DNA technology. Of the many restriction enzymes that do chop DNA precisely, some cut both strands of the DNA double helix at the same place, so that 'blunt-ended' fragments are produced (Figure 27). Others cut the two strands in a staggered manner and create products containing 'sticky' ends. Enzymes that produce sticky-ended DNA fragments are especially useful for manipulating DNA because the sticky ends produced by one restriction enzyme are the same regardless of the DNA molecule that it cleaves. Sticky ends recognise each other and, as their name implies, they stick to each other (Figure 28). This allows two DNA fragments, obtained by cutting different DNA molecules with the same restriction enzyme, to be joined together to produce a hybrid,

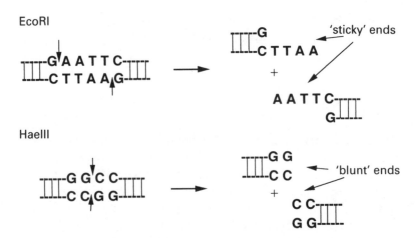

Figure 27. Restriction enzymes cleave DNA to produce 'sticky' or 'blunt' ends. Sticky ends are produced when the enzyme cuts DNA at a site away from the centre of the recognised sequence, as with the enzyme, EcoRI (top). When the enzyme cuts DNA at the middle of the recognition sequence blunt ended fragments result, as with the enzyme, HaeIII (bottom).

or **recombinant DNA** molecule. For example, a human DNA segment could be mixed with a phage DNA segment to produce a human–phage recombinant DNA molecule. This procedure has, indeed, been used to insert human genes into a phage's DNA encyclopaedia, so that the human gene can be isolated and grown as if it were part of a phage.

Many restriction enzymes cut DNA within the recognition sequence of DNA letters that they specifically identify. Some, however, recognise a specific sequence but cut DNA outside of this sequence. Nevertheless, most recognition sequences have symmetry. It is well established that each of the four DNA 'letters', A, G, C and T, on one strand of the DNA double helix is always associated with one of the other letters on the opposite strand of the double helix. Thus, T is always opposite A, and G is always opposite C. The two strands have direction, so that the sequence, AACGT, for example, is not the same as TGCAA: the left-hand end of a DNA molecule is different from the right-

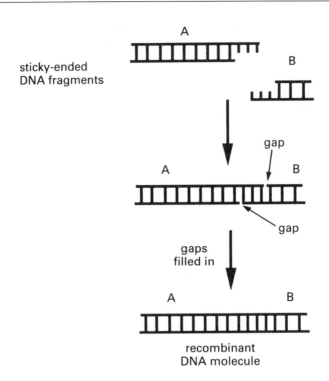

Figure 28. Production of 'sticky ended' fragments of DNA
by restriction enzymes is useful for joining two DNA molecules
together. Here, different fragments of DNA (A and B) produced by digesting two separate DNA molecules with the same
restriction enzyme stick together and the gaps between their
strands can be closed when chemical bonds form between the
two 'free' ends so as to join them together. This produces
a hybrid (recombinant) DNA molecule containing both
sequences, A and B.

hand end, and the proper way of writing these two sequences is:
5'AACGT3' and 5'TGCAA3'. (The 5' and 3' refer to specific
chemical features of the DNA molecule, which make the left
and right ends different.) In other words, 3'TGCAA5', but not
5'TGCAA3', is the same as 5'AACGT3'. If the two strands of
the DNA double helix within the recognition sequences of many
restriction enzymes are examined, the sequence of one strand is

identical to the sequence of the other strand. For example, EcoRI recognises and cuts each DNA strand within the sequence, 5'GAATTC3'. In a DNA double helix, the opposite strand to this sequence is also 5'GAATTC3', since A always pairs with T and G with C, and the 5'-end of one strand is always opposite the 3'-end of the other strand, and *vice versa*. This type of symmetry is called two-fold rotational symmetry: when the double helix is rotated through 180 degrees, it is identical to the starting

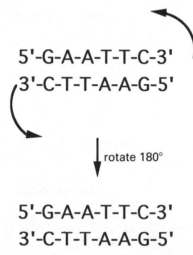

5'-G-A-A-T-T-C-3'
3'-C-T-T-A-A-G-5'

rotate 180°

5'-G-A-A-T-T-C-3'
3'-C-T-T-A-A-G-5'

Figure 29.   Restriction enzymes frequently bind to and cut within DNA sequences that possess symmetry. The sequence recognised on one strand of the DNA double helix is identical to the sequence on the opposite DNA strand. This can be seen here with the EcoRI recognition sequence: when the sequence is turned through 180 degrees, the sequence remains the same.

molecule (Figure 29). These symmetrical sequences resemble palindromic words, such as 'civic' and 'refer', which read the same in both directions, and so the DNA sequences are called palindromic DNA sequences. The symmetry reflects the way in which the restriction enzyme binds to the DNA and cuts it on opposite strands at equivalent sites.

In bacteria each restriction enzyme has a corresponding DNA methylase that modifies the restriction enzyme's recognition sequence and prevents the DNA from being cut. For example, EcoRI, which cleaves each DNA strand between the G and the A of the sequence, GAATTC, exists in *Escherichia coli* along with a DNA methylase that modifies the second A of this sequence. When this A is modified, EcoRI will no longer cut the DNA. In this way, all of the GAATTC sequences that occur in the bacterium's DNA encyclopaedia are modified by the DNA methylase, which adds a chemical group to the A, and so the bacterium's EcoRI restriction enzyme does not cut its own DNA. Foreign DNA, either in phage or from other sources, lacks the modification and so any GAATTC sequences are cut. In order to protect the foreign DNA from being chopped by the EcoRI, the second A of its GAATTC sequences must be modified by the DNA methylase.

Most restriction enzymes recognise sequences that are four or six letters long, but some recognise sequences containing more letters. Other restriction enzymes recognise a specific DNA sequence but cleave outside of this sequence at a precise number of letters from the end of the sequence. Restriction enzymes have been isolated from over a hundred different microorganisms; the diversity of sequences they recognise and cleave provide scientists with molecular scissors tailored to different needs. DNA molecules can be chopped into pieces, joined together, inserted into the DNA encyclopaedias of animal, plant and microbial cells, and now it is possible to create **transgenic** animals and plants, which contain and express foreign genes that normally do not exist in those organisms. Plants, for example, can be made to produce animal antibodies by inserting antibody genes into their DNA encyclopaedias. The potential for production of medically and agriculturally useful products is staggering. The Age of Genetic Engineering is with us, and it will not only provide useful products: it will also open up the possibility of treating, perhaps curing, genetic diseases such as cystic fibrosis and muscular dystrophy. Restriction enzymes began with humble origins, when scientists were trying to understand how bacteria protected themselves from invading phage. They have reached a stage at which they have changed the world through their contribution to recombinant DNA technology.

17

DNA, the molecular detective

In 1983, a teenage girl was raped and strangled in Enderby, Leicester, in England. Three years later, and only a mile from the scene of the crime, a fifteen-year-old girl was found similarly raped and murdered. Leicestershire police believed that both girls were victims of the same criminal, but had no hard evidence to support this theory. At that time, in 1986, a research group led by Professor Alec Jeffreys (b. 1950) in Leicester University's Biochemistry Department, had developed an amazing new scientific method for identifying and discriminating between individual human beings. This technique, called **DNA fingerprinting**, was so accurate that the chances of any two people (with the exception of identical twins) in the entire world having the same DNA fingerprint were virtually zero. Unlike conventional fingerprinting, a criminal's hands were not required: DNA fingerprinting could be carried out on minute quantities of body fluids such as blood or semen. Even dried bloodstains made years earlier could be used to obtain a DNA fingerprint that might identify the person who spilt the blood.

Traces of semen from the Enderby murderer or murderers were found on the victims and Professor Jeffreys was approached to try out his new DNA fingerprinting method on these samples in order to ascertain whether or not the criminal was the same for both murders. The test proved conclusive: semen from one victim had an identical DNA fingerprint to that from the other victim. Both murder victims had been raped by the same man, and police suspicions were confirmed.

A seventeen-year-old youth from the area in which the killings took place became a major suspect and even confessed to the crimes. However, Jeffreys carried out a DNA fingerprint analysis of him and found that it did not match the DNA fingerprint obtained from the semen samples of the true rapist. It was clear that the suspect youth had not carried out the crimes, despite his confession. A major and intensive hunt for the real killer was instigated. The police did have an important piece of evidence: the criminal's DNA fingerprint.

In early 1987, the police launched a mass screening of the local population, in which about 5000 blood samples were collected from different individuals and their DNA fingerprints found. Unfortunately, none of these matched the DNA fingerprint determined from the semen samples taken from the rape victims. Yet the police strongly suspected that a local person had carried out the killings. Not long afterwards a man informed police about a conversation he had overheard between two workmates. One had said to another that a baker called Colin Pitchfork had persuaded him to give blood during the mass DNA fingerprint screening programme carried out by the police. When Pitchfork was approached by police and his DNA fingerprint was analysed, it did, indeed, match that found in the semen samples from the victims of the rapes. In 1987 Pitchfork became the first murderer to be convicted using DNA fingerprinting as key evidence.

DNA fingerprinting has since become a world-wide forensic technique and new variations of the original method are improving its power to help solve crimes. For example, a person now can be identified many years after he or she has died by examining DNA from bone tissue, and DNA isolated from the root of a single hair can be analysed. DNA fingerprinting is frequently used to solve paternity cases: the real father of a child can be unambiguously identified and any male who is not the real father can conclusively be shown not to be so. Immigration disputes are also being solved by Jeffreys' method. For example, in many cases where a man claims to be the father of a family, he is suspected of being, in reality, the real father's brother. DNA fingerprinting can discriminate between these two possibilities.

Previous forensic tests on biological samples were based largely on blood group analysis, which can be used only if a blood sample is available and cannot be used on other tissues or body fluids. The molecules that carry blood groups are unstable, so this analysis cannot be carried out on aged specimens. Blood groups are only moderately variable from one person to another, so they frequently provide ambiguous results. DNA fingerprinting does not have these drawbacks. Whilst previous methods allowed a person to be eliminated as a suspect, they could never by themselves prove beyond reasonable doubt that a suspect was the true criminal. Provided that the material to be examined is not contaminated with DNA from other human sources, that DNA is available from the suspect, and that the procedure is carried out properly, guilt or innocence of a suspect can be established.

One of the remarkable facts about DNA fingerprinting is that it was discovered quite by chance. Alec Jeffreys, the discoverer, had no idea when he was carrying out his biochemical research, that he would eventually stumble upon a method that would revolutionise forensic science and be so directly useful to society. Here is another case of basic research producing unforeseen benefits: DNA in a laboratory test-tube had become a molecular Sherlock Holmes!

In order that DNA fingerprinting can be appreciated properly, some further understanding of the nature of genes and DNA is needed.

DNA, genes and proteins

It is obvious that each individual (with the exception of identical twins) has his or her own unique set of physical features. Otherwise, how would we recognise each other? The physical appearance of a human being or any other living organism is called the **phenotype** of that organism. Phenotype is determined largely by the interaction between hundreds of thousands of molecules in our bodies. In particular, proteins have a major role in determining phenotype because they are constituents of many structural components of living cells and they carry out many of the chemical processes, such as conversion of food energy to muscular activity, that occur in our bodies. Eye colour, hair colour and texture, body height and build are all determined to a large degree by proteins.

Where do these proteins come from? Most of them are made by our body cells: they are built up of chemicals called **amino acids**, which come from our food or which our cells make themselves from other food components such as sugars. There are tens of thousands of different proteins in the human body, each one made up of many hundreds of amino acids arranged in a particular sequence. This means that each protein is distinct from any other and is suited to carry out a precise function in the body. Insulin is a protein that controls blood sugar levels; antibodies are proteins that are involved in attacking and destroying foreign bacteria and other invaders of the body; keratin is a protein that forms a major part of the structure of hair and finger nails; and haemoglobin is a protein that carries oxygen to our body cells so that they can thrive.

The key to understanding how proteins are produced in precise ways by different cells lies in the **genes**. Genes code for proteins and they therefore determine much of our phenotype. Each protein in the body is coded for by its own unique gene. Almost every cell in the body, be it a brain, liver, kidney, skin or lung cell, contains all of the genes needed to make a whole organism. Thus, a liver cell contains genes that code for proteins that are found specifically in the brain in addition to those that code for proteins found only in liver cells or those found in both liver and brain cells. The difference, therefore, between a liver and a brain cell, is determined by which particular set of genes in those cells is used to make proteins. A liver cell has been programmed to make only those proteins that are required by a liver cell: the genes for these proteins are 'switched on'. Brain-specific proteins are not needed by a liver cell, even though it has the genes that code for them. The genes for brain-specific proteins are therefore 'switched off' in liver cells and so brain-specific proteins are not made by liver cells. It is this selective expression of different sets of proteins that produces different cells and tissues in the body.

Genes are passed to us from our parents in the spermatozoon and egg that started our development off. Half of our genes come from our mother, the other half from our father. Much of our individual uniqueness comes from the combination of genes we inherit from each parent. There are millions of possible combinations in which we can inherit genes from each parent, and that is why even our brothers and sisters differ substantially in appearance from us. Yet they also have many genes in common with us and hence bear some resemblance to us and to one another.

Genes are nothing mystical even though they are a remarkably complex and beautiful manifestation of nature. They are made up of DNA and the mechanism by which DNA in a cell is 'read' to produce the proteins that determine a person's phenotype is now understood in large measure. This understanding of DNA and how it codes for proteins constitutes one of humankind's greatest achievements of the twentieth century.

The sequence of amino acids in a protein is unique to that protein and gives the protein its structural and functional properties. DNA is a simple four-letter alphabet that determines what the amino acid sequence of a protein is. The four letters, A (adenine), G (guanine), C (cytosine) and T (thymine), are chemically distinct structures (bases) joined together in long chains. The gene for a particular protein can be read in terms of the four different letters to give the amino acid sequence of the protein. For example, the insulin gene consists of a stretch of DNA that contains many thousands of the four letters arranged in a very specific way: when a cell makes insulin it reads the information in the insulin gene and makes the insulin protein.

DNA is, in essence, a remarkable molecular encyclopaedia which living cells read in order to make their proteins so that they can build up their structures and carry out life's processes. Each human cell carries a DNA encyclopaedia that describes every protein in the whole body, but only a fraction of the 'paragraphs', or genes, are needed by a particular type of cell. So a brain cell, for instance, reads only the paragraphs it needs; that is to say, it produces only the proteins it requires for its proper function.

The DNA encyclopaedia of an organism is called its **genome**. The human genome contains about one hundred thousand different genes. If the DNA from a single human cell was isolated and untangled, it would stretch to almost two metres (six and a half feet) long. Since the average adult contains about a million million cells, if the DNA was extracted from every cell in the body it would stretch from the Earth to the Moon and back several thousand times.

The past twenty years have seen gigantic strides in our understanding of genes and our ability to artificially manipulate them in the laboratory. DNA coding for almost any part of the human genome can now be isolated and obtained in a small test-tube. We are in the realms of genetic engineering, which one day may allow genetic illnesses such as cystic fibrosis, Down's syndrome and muscular dystrophy to be treated by manipulating DNA. At any rate, the great progress made in DNA manipulation has allowed defective genes in diseases such as cystic fibrosis and muscular dystrophy to be studied. Such studies are revealing clues that might lead to better treatment of these and other unpleasant ailments.

It turns out that a large fraction of DNA in a human cell actually does not code for proteins. Some biologists call this DNA 'junk', although it is now clear that at least some of it has important functions. Although there is some variability in protein-coding genes from one person to another, 'junk' or non-coding DNA is vastly more variable and it is this that forms the basis of DNA fingerprinting.

DNA fingerprinting arose from basic studies of human DNA and how it varies between individual people. Its forensic applications were not even a remote goal of this research; rather, they were one of its unexpected by-products. Such is the way that scientific research progresses: nobody can be sure what treasures it has in store.

Discovery of DNA fingerprints

Alec Jeffreys was born in Oxford, England, in 1950. He became interested in the molecules that make up living cells at an early age and took a biochemistry degree at Oxford University. In the early 1970s, scientists had developed techniques for studying how DNA codes for proteins and why it is that different types of cell produce different sets of proteins. Jeffreys was fascinated with this area of research and did his doctoral research on the subject.

Towards the end of the 1970s and in the early 1980s, scientists had progressed to a stage at which genes were being isolated. It was also possible to determine the sequence of the four letters (A, G, C and T) along a given DNA molecule and therefore to 'read' the proteins encoded by them. Like many scientists, Jeffreys believed that isolation of DNA molecules and determination of their sequences held the key to understanding how cells control which set of genes to translate into proteins, how genes had evolved, and how inherited diseases such as cystic fibrosis and muscular dystrophy, which involve abnormal DNA sequences, might be better understood. He therefore joined a team of scientists in the Netherlands who were isolating and studying the DNA that codes for globin, a protein constituent of haemoglobin. Haemoglobin gives blood its red colour and plays a vital role in carrying oxygen from the lungs to other parts of the body. The gene that codes for globin was one of the first genes ever to be isolated in a test-tube.

Whilst he was working with Dick Flavell (b. 1943) in the Netherlands, Jeffreys discovered that genes were not as simple as many people thought. Instead of being continuous, the globin gene was found to be interrupted by DNA sequences that were apparently meaningless. These interruptions, which are called **introns,** do not code for any part of the globin protein molecule, yet they occur within the gene that codes for globin. It is as if the paragraphs in the DNA encyclopaedia have nonsensical stretches of words interspersed within many of its meaningful sentences. Introns have since been found in many genes and there is some evidence that, far from being gibberish, they are involved in regulating the levels of protein produced from a particular gene.

Jeffreys' studies of the globin gene inspired his curiosity and his desire to learn more about DNA sequences and how they varied between different individuals and different species of organisms. He returned to Britain to a lectureship at Leicester University, where he began to study another gene, this time one that codes for a protein called myoglobin. Myoglobin is similar to haemoglobin, but unlike haemoglobin it is not present in the blood and instead is a constituent of muscle cells. Myoglobin helps muscle cells to receive oxygen from the blood for use in muscular activity. At the same time, the team of scientists Jeffreys was leading was also working on human DNA sequences called **minisatellites.** Minisatellites, like introns, are DNA sequences that do not code for proteins; they occur in thousands of copies in each human cell and their functions are not yet fully understood.

The two research projects on myoglobin and minisatellites converged when Jeffreys and his colleagues found a minisatellite DNA sequence in the human myoglobin gene. This minisatellite was present in one of the introns of the myoglobin gene.

Jeffreys was interested in this minisatellite in the same way that he was interested in DNA sequences in general and how they were arranged in the human genome. Were minisatellites found elsewhere in the genome, within genes coding for proteins other than myoglobin? Jeffreys believed that if this were the case, minisatellites might be used as 'markers' for different gene regions and that they might allow detection of gene regions that were abnormal in various inherited ailments, in which abnormal DNA sequences code for faulty proteins, and hence diseased phenotypes.

When Jeffreys' team looked for the myoglobin minisatellite DNA sequence elsewhere in the human genome they found, indeed, that similar sequences did occur at many other sites. What was surprising, however, was that different people had different patterns of minisatellites in their genomes. Variability in the lengths of minisatellite DNA stretches was so great between different people that Jeffreys realised he had stumbled upon a DNA sequence that allowed one human being to be distinguished from another – a DNA fingerprint. He also discovered that related people have some, but not all, features of each others' DNA fingerprints in common. A child's DNA fingerprint, for example, could be constructed from part of each of its mother's and father's DNA fingerprints. Figure 30 shows a representation of how DNA fingerprints vary between members of a family.

After Jeffreys and his colleagues had published their scientific paper demonstrating that DNA fingerprints exist, the national newspapers in Britain reported the find. Jeffreys was, as a result of this publicity, contacted by a lawyer who was having problems solving an immigration case. A man had been refused entry into Britain although he claimed to be a legitimate member of a family that was already living there. He and his lawyer had failed to convince the authorities of the authenticity of his relationship with the family. Jeffreys and his team carried out DNA fingerprinting of this man and members of his alleged family and proved beyond reasonable doubt that he was who he claimed to be.

The newspapers followed up this story with great zest and Jeffreys was overwhelmed with requests for DNA fingerprints to be carried out on other immigrants. Eventually, at the end of 1985, a company was set up to carry out the procedure. Forensic laboratories in Britain and other countries soon developed the necessary technology and DNA fingerprinting is now used world-wide to solve crimes as well as paternity and immigration disputes.

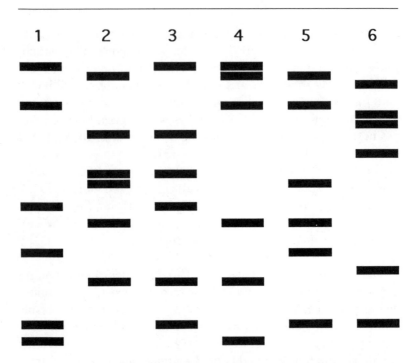

Figure 30.   DNA fingerprints amongst family members. Lane 1 shows the mother's DNA fingerprint; lane 2 shows the father's DNA fingerprint; and lanes 3, 4 and 5 show three of their children's fingerprints. Note that each child receives bands from each parent and all of a child's bands are found in one parent or the other. The DNA fingerprint in lane 6 comes from someone who is not a member of the family: the bands are quite different from those seen in the parents or children.

The technique of DNA fingerprinting

The starting point for making a DNA fingerprint is to obtain a sample containing DNA of the person whose fingerprint is to be determined. Nowadays, DNA fingerprinting or a modification of the procedure can be carried out on a whole range of substances, including blood, semen, saliva, hair and bone. DNA is tough and has even been recovered in small amounts from a two thousand five hundred-year-old Egyptian mummy and from five thousand five hundred-year-old bones. Jeffreys and his team identified a murder victim by extracting DNA from the skeleton and comparing it to the DNA of the victim's putative parents. They also found that good quality DNA could be isolated from four-year-old bloodstains on cotton cloth and used to make fingerprints.

Recently it has become possible to use DNA fingerprinting and related methods on minute samples. A single hair root contains about five thousand cells and this is enough to give a clearcut DNA pattern. Indeed, it is now possible to detect the DNA of a single cell. This has been due to another procedure called 'polymerase chain reaction' or PCR, which allows amplification of DNA from a single cell into many thousands of copies that can then be examined using Jeffreys' method. Forensic medicine will never be the same again: DNA recovered from the scene of a crime even in tiny quantities will identify victims and criminals alike on an unprecedented scale.

Once DNA has been isolated from the sample of body fluid or tissue under study, its quality and quantity are assessed to make sure that it is in sufficiently good condition for DNA fingerprinting to be carried out. Figure 31 illustrates the subsequent methodology involved in obtaining a fingerprint from this DNA. The DNA is broken down into smaller fragments of defined sequence using restriction enzymes, proteins that cut DNA at precise sites (Chapter 16). This allows the different regions of a person's genome that contain minisatellites to be separated from each other. Some minisatellites will be found in very large DNA fragments, others in small fragments; and some fragments produced by restriction enzymes will not contain minisatellite sequences.

The DNA fragments are then separated according to their size using a procedure called **electrophoresis**. In this method, large DNA fragments move more slowly than smaller ones under the influence of an electric current. Once separated, the fragments containing minisatellite DNA need to be specified, since they represent only a small proportion of the total number of DNA fragments obtained. A 'probe' that detects minisatellite DNA is used for this purpose and the final DNA fingerprint consists of a number of bands (differently sized DNA fragments containing

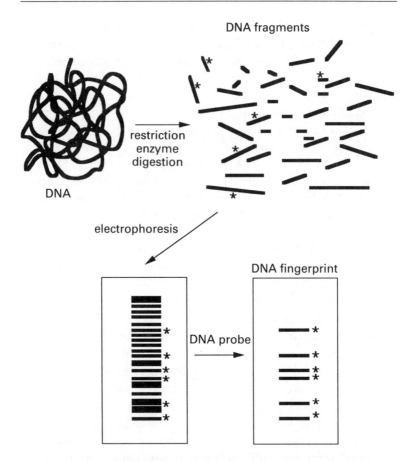

Figure 31.   DNA fingerprinting. DNA is extracted from the sample (blood, semen, etc.) to be fingerprinted and fragmented into defined pieces. The DNA fragments are separated by the process of electrophoresis on the basis of their size, then studied with a DNA probe that detects fragments containing minisatellite DNA sequences: the final step produces a fingerprint showing the DNA fragments containing minisatellite DNA. Fragments containing 'fingerprint' sequences are marked with a star.

minisatellite DNA) arranged in a precise pattern that is a characteristic of one person.

DNA fingerprints have been compared with the bar code on

supermarket groceries. Each item has a unique bar code that differs from any other item and allows the automatic cash register to identify whether it is a can of beans, a bag of flour or something else. DNA fingerprints identify individual people with similar accuracy and simplicity.

Applications of DNA fingerprinting to areas other than forensic science abound, and new uses of the method are frequently being found. Minisatellite-like DNA has been found in other animals and even in plants, and DNA fingerprinting can be used to identify stolen animals, to verify pedigrees in animal breeding disputes, and to characterise plants produced during breeding studies. Jeffreys' DNA method was used, for example, to confirm the pedigree of a championship-winning dog in a British dog show.

Because transplanted organs are derived from a donor whose DNA fingerprint differs from the recipient, DNA analysis can be used to study the progress of organ transplants. DNA fingerprinting of bone marrow cells taken from a leukaemia patient who has received a bone marrow transplant can, for example, allow an assessment of how well the transplanted cells have established themselves in their new human body.

In the late 1980s, a body was found in Brazil that was said to be that of Josef Mengele, the Nazi doctor, who was known as the 'Angel of Death' by victims of the Auschwitz concentration camp during the Second World War. Evidence was examined and experts came to the conclusion that the body did, indeed, belong to Mengele, although some doubts remained. Subsequently, DNA fingerprinting was carried out on the body in conjunction with DNA fingerprints of close relatives of Mengele who are alive today. The results confirmed beyond reasonable doubt that the body did belong to the atrocious Nazi doctor.

DNA fingerprinting may also be useful for identifying accident victims, for example those killed in explosions or plane crashes, who cannot easily be recognised by other criteria.

What started as a pure scientific pursuit for Alec Jeffreys became a crime-solving weapon of great power, and its uses have extended into many other areas of life. Even when his team of researchers had established the individual specificity of DNA fingerprints, Jeffreys had his doubts as to whether the legal profession would accept his molecular evidence in court: 'At the time [of the discovery of DNA fingerprints] we suspected that ... major problems would be encountered as DNA evidence proceeded from the research laboratory to the court room. Subsequent history showed that we were unduly pessimistic.'

18
Magic bullets

T he second half of the twentieth century has seen fantastic
strides in recombinant DNA technology that will undoubtedly bear rich fruits of new drugs and agricultural products, as well as a greater understanding, prevention, diagnosis and
treatment of many diseases. Another monumental step forward in
research on diseases was made in 1975, in the field of immunology, the study of the defence systems that living organisms use
to fight off infectious organisms. This breakthrough provided
powerful tools – **monoclonal antibodies** – for biological, medical and biochemical research. Monoclonal antibodies have made
a tremendous contribution to science. Their medical applications include diagnosis of cancers and other diseases, pregnancy
testing and tissue typing, and many more benefits, including
their use as anti-cancer drugs, may well be forthcoming in the
future.

In order to appreciate what monoclonal antibodies are and why they are so important, it is necessary to understand some aspects of the immune systems of animals. Without the immune system it is unlikely that most of us would survive long after birth under normal conditions: we would soon succumb to infectious diseases. The drastic effects of the loss of the immune function can easily be seen in patients who suffer from acquired immunodeficiency syndrome (AIDS), which involves impairment of certain components (T-cells) of the immune system. These patients are prone to developing infections that people without AIDS rarely, if ever, contract. For example, a major complication of AIDS is pneumocystosis, a disease of the lung that is caused by a microbe which occurs almost everywhere in the environment. We all encounter this microbe every day and breathe it into our lungs, but it does not cause us any harm because we developed immunity to pneumocystosis when we were young. Our immune system fights off the microbe successfully every time we come into contact with it. With AIDS patients, however, the defence system is defective because the AIDS virus damages the T-cells of their immune system. Because T-cells are an important part of the defence against foreign microbes, including the one that causes pneumocystosis, AIDS patients no longer have immunity. Consequently, they easily become infected with the pneumocystosis microbe. AIDS patients are prone to several other diseases that are caused by microbes which we encounter almost daily, and are normally harmless, because we all develop immunity to them from an early age. Such infections are said to be **opportunistic** ones.

In addition to protecting us from normally harmless microbes that we encounter in our everyday environment, the immune system fights off many infectious organisms that cause us harm, allowing us to recover from them. It is also the immune system that allows us to be vaccinated against diseases so that we do not become infected with them in the future. The immune system is important for many other reasons, and knowledge of its complex mechanisms is essential for better treatment and prevention of infectious and other diseases, and also for improving survival of transplanted organs in patients who receive them. Transplant rejection occurs because the immune system of one person 'knows' the difference between organs of that person and those belonging to another person.

Cells such as the T-cells, which are impaired in AIDS patients, constitute one component of the immune system. Another part of the system consists of non-cellular substances, including proteins called **antibodies**. Antibodies are highly versatile molecules that specifically recognise and bind to foreign substances, such as bacteria or proteins. Generally, each antibody binds to only one single molecule (or part of a molecule) on the foreign substance, and it will not bind to a different substance. For example, antibodies that bind to a particular protein will fail to bind to another, different protein, but there may be antibodies that bind to the second protein and not to the first. More particularly, antibodies that bind to one part of a particular protein will not usually bind to another region of that protein. In other words, antibodies are highly specific. The substance (which may be a protein or another type of molecule) to which an antibody binds is called an **antigen** and therefore, generally speaking, each antibody recognises and binds to only one type of antigen or even to only one region (an **antigenic determinant**) of a particular antigen.

When a foreign organism or other substance enters our bodies, we frequently produce antibodies to it. These antibodies appear in our bloodstream and can readily be detected using simple biochemical techniques. The binding of antibodies to antigens starts off a process in the body that eventually counteracts and eliminates the foreign substance. If we produce antibodies against a bacterium, there may be many different antibodies in our bloodstream against that bacterium. Some of the antibodies will be against one particular antigen in the bacterium, other antibodies will be against different antigens, and several antibodies may recognise different antigenic determinants of a single antigen of the bacterium. There may even be different antibodies that bind to the same antigenic determinant, some fitting better than others, just as different pairs of shoes can fit the same pair of feet. Even though we may never have encountered the bacterium previously in our lives, we are likely to produce many antibodies that recognise its antigens specifically. Perhaps even more surprising, antibodies can be produced that bind specifically to man-made substances such as small synthetic molecules, which neither occur naturally nor are likely ever to enter an animal under normal circumstances. Antibodies that recognise these chemical antigens, and will not recognise any other known substance, can readily be produced in laboratory animals.

It has been estimated that the number of different antigens that the antibodies of one human being can potentially recognise is well over a million, and possibly tens of millions. In other words, our immune system has the capacity to produce millions of different antibodies, each of which recognises its own particular antigen. How can this be? What wonderful mechanism accounts for this enormous versatility and specificity, which allows an animal not only to recognise millions of foreign substances as alien, but also to be able to discriminate between each of these substances? It is as incredible as if one person could recognise, and be able to tell the difference between, each of more than a million other human beings.

The search for the explanation of the complexity of the antibody response of animals to foreign substances led, somewhat fortuitously, to the development, in 1975, of a method for monoclonal antibody production. The two scientists who made this discovery were Georges Kohler (b. 1946) and Cesar Milstein (b. 1927), who were working at Cambridge University, England. In 1984 they were awarded the Nobel Prize for Physiology or Medicine for this important step forward. Even before Kohler joined Milstein's laboratory, Milstein had been studying antibody production. Milstein was born the son of a Jewish immigrant in Argentina and trained there as an experimental scientist. He went to Britain in 1958, returned to Argentina in 1961, and finally went back to Britain in 1963. At Cambridge University he and his colleagues became established members of a growing group of scientists around the world who were trying to understand why it is that one animal can produce millions of antibodies with different antigen specificities. This problem goes back a long way in the history of the study of antibodies.

The history of antibodies

The work on antibody diversity has much of its historical beginnings in the development of vaccines by Jenner and Pasteur (Chapter 12). By the end of the nineteenth century it was clear that not only are animals capable of defending themselves against foreign microbes, but also immunity to many microbial diseases can be artificially induced by injecting animals with dead or inactive forms of the disease-causing microbes. The advent of vaccine development by Pasteur led many scientists to investigate seriously the mechanisms of immunity to various diseases.

A major development in immunology took place in 1890, when the German bacteriologist, Emil von Behring (1854–1917) and the Japanese bacteriologist, Shibasaburo Kitasato (1852–1931), demonstrated that immunity to tetanus was due to the presence of substances in the bloodstream. It was known at the time that tetanus bacteria produce a poisonous substance that causes many of the symptoms of the disease and that the poison could be obtained from the broth in which these bacteria are grown. Von Behring and Kitasato injected rabbits with sublethal doses of the tetanus poison and discovered that this produced immunity in the rabbits: they failed to succumb to subsequent injections of live tetanus microbes that killed non-immunised rabbits. Von Behring and Kitasato then took some blood from these immunised rabbits and separated out the blood cells so that they were left with the non-cellular part of the blood, which is called **serum**. This serum was injected into mice, which were then challenged with infectious tetanus microbes. The mice failed to develop tetanus and it was clear that something, which was called an **antitoxin,** in the serum of the immunised rabbits was protecting the mice (and the immunised rabbits) from the tetanus poison. This great discovery opened the way for **serum therapy**, in which serum from immunised animals is transferred to humans infected with a disease; the antitoxin in the transferred serum provides temporary protection from the spread of the disease. Von Behring subsequently applied serum therapy to diphtheria, which was a major killer of young children in the late nineteenth century.

Serum from horses immunised with the poison produced by diphtheria bacteria proved to be effective at protecting children temporarily from diphtheria, and also was valuable for treating children already infected with diphtheria. By 1894, this antidote to diphtheria was commercially available and was considered to be a major step forward in the treatment of human disease.

We now know that the antitoxins present in these protective sera were antibodies that bound to and inactivated the poisons of tetanus and diphtheria microbes. In 1895 antibodies were detected in serum in a different way. The Belgian scientist, Jules Bordet (1870–1961), discovered that the red blood corpuscles of one animal clumped together when they were incubated in the presence of serum from another species of animal. Something in the serum of one species was causing the red cells of another species to aggregate. Five years later, Karl Landsteiner (1868–1943), the Austrian immunologist, showed that serum from one human being can cause clumping of red blood cells from another person. Landsteiner found that red cell clumping patterns could be used to classify human blood into three different major types, which he called A, B and O types; a fourth type (AB) was later added. He had discovered blood groups, which opened the way for much more successful blood transfusions, since an individual's blood group could now be determined, ensuring that any transfused blood they received came from a donor with a compatible blood group. The serum factors that cause red blood cell clumping were called **agglutinins**; we now know that they are antibodies.

Paul Ehrlich (1854–1915), the great German scientist, was highly influential in immunological thought at the beginning of the twentieth century: his theories about antibody diversity and specificity dominated the field. Ehrlich compared the high degree of specificity of antibodies for their antigens to a lock and a key: antibodies (keys) occur in many shapes and each antibody will fit only one antigen (lock). Ehrlich proposed an explanation for antibody diversity which he called the **side-chain hypothesis**. His idea was that each cell in the body that produces antibodies has a number of antibody molecules on its surface which recognise different antigens. These antibodies are present before the body encounters the antigen. When an antigen enters the body, it binds to its specific antibody on the surface of an anti-body-producing cell and this triggers the cell to produce more of that particular antibody. In other words, antibody-producing cells express antibodies to more than one type of antigen, and an encounter with one of these antigens causes the particular antibody to that antigen to be produced in large amounts. This remarkable idea is a credit to Ehrlich's genius, since it is very near to the truth and yet was proposed at a time when extremely little detailed knowledge was available about antibodies. We now know that antibodies that bind to particular antigens already exist before a person encounters the antigen, as Ehrlich believed, although it is now clear that each antibody-producing immune cell makes antibody to only one antigen, and not to several.

Ehrlich's theory eventually gave way to two major ideas that were propounded to account for the high degree of antibody diversity. These were the **instructive theory** and the **clonal selection theory** (Figure 32). The instructive theory proposed that all antibody molecules (which are proteins) are identical – they have the same amino acid sequence (Chapter 17), but they can fold in millions of different ways. According to this theory, when an antigen encounters an antibody, the antibody folds around the antigen and assumes a close fit with it: this triggers production of more antibody folded in this way. In the instructive theory, the folded antibody to a particular antigen does not exist in the body before an unfolded antibody encounters the antigen. On the other hand, the clonal selection theory stated that the immune system contains cells that produce millions of different antibodies with distinct antigen-binding specificities: each cell produces an antibody with a single specificity, and therefore there are millions of different antibody-producing cells. In the clonal selection theory, when an antigen enters the body it binds to antibodies on the surface of those cells that fit the antigen well. This triggers production of more of this particular antibody, or of more of the cells that produce this antibody. According to the clonal selection theory, antibodies to particular antigens exist already before the body sees the antigens.

INSTRUCTIVE THEORY

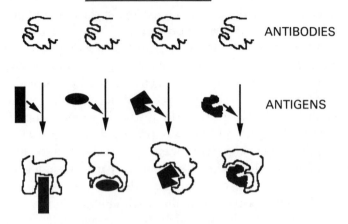

CLONAL SELECTION THEORY

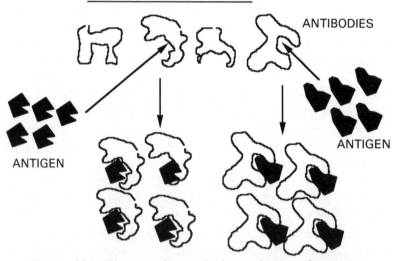

Figure 32. Theories for antibody production. Instructive theories held that all antibody molecules were identical and that they folded into specific structures when they came into contact with antigens (shown in black here). Clonal selection theories stated that antibodies were different from each other before they came into contact with antigens, and that antigens merely selected the antibodies they could bind: this triggered production of further antibody molecules of the type selected.

In the 1960s the English biochemist, Rodney Porter (1917–1985) and the US biochemist, Gerald Edelman (b. 1929), established the molecular structure of antibodies, which they dis-

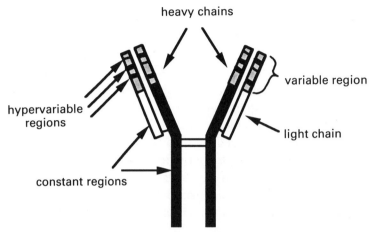

Figure 33. Antibodies are Y-shaped molecules consisting of four chains: two heavy and two light chains. Each chain contains a variable and a constant region. The variable regions give each antibody its unique antigen-binding properties. Variable regions also contain hypervariable regions, which are especially important for antigen-binding specificity.

covered to be Y-shaped molecules (Figure 33). Porter and Edelman were awarded the 1972 Nobel Prize in Physiology or Medicine for this great step forward in our understanding of the immune system. The Y-shaped antibody molecule contains regions that are essentially identical from one antibody to another: these **constant regions** appear to be the same regardless of the antigen that a particular antibody recognises. Other parts of the molecule do vary from one antibody to another, and these **variable regions** occur in the portion of the antibody molecule that binds to its specific antigen. This means that antibodies are all Y-shaped molecules with many features in common, but they differ from each other at particular sites and it is these regions that determine which antigen they will recognise. The fact that antibodies that bind to different antigens are not entirely identical (their amino acid sequences differ) was contrary to the instructive theory for antibody diversity, which required that they had identical amino acid sequences, and so the clonal selection theory eventually became the established explanation for antibody diversity.

Once the clonal selection theory was accepted, it became necessary to explain how the DNA encyclopaedia, or genome (Chapter 17), of an animal could encode the many different protein molecules necessary to account for the millions of different antibodies. If these antibodies already exist before an animal encounters the corresponding antigens, does this mean that each antibody protein molecule is encoded by its own gene in the animal's DNA encyclopaedia? This would require there to be millions of genes coding for antibodies alone, and it would mean that antibody genes would be the most abundant genes in an animal's genome. Many scientists thought it unlikely that there was enough information in the DNA encyclopaedia of a human being or any other animal to include millions, or possibly even tens of millions, of different antibody molecules.

Two theories arose. In the first, the millions of distinct antibodies were all encoded by the DNA encyclopaedia; in other words, there *were* millions of antibody genes in the genome and these were passed from parents to offspring in the same way that genes for other proteins were inherited. In the second theory, there were only a few genes encoding antibodies in the DNA encyclopaedia, and antibody diversity was created by special biochemical mechanisms that occur in the immune cells, which are known as **B-cells**, that produce antibodies. Thus, each B-cell would produce a single type of antibody that recognised a particular antigen and the amino acid sequence of each antibody would then be encoded by the DNA of the particular B-cell that produced it. In this case, the DNA sequence for an antibody produced by a particular B-cell had become specifically varied in that cell. In other B-cells the antibody gene had varied in a different way, producing antibodies with different antigen specificities. In the non-B-cells of the body, such as liver or brain cells, which do not produce antibodies, there were only a few antibody genes and they were the same in all cells of that tissue.

The process that created antibody gene diversity was said to be specific for antibody-producing B-cells.

It turns out that the two ideas concerning the genes that encode antibodies each have some truth in them. In the DNA encyclopaedia, there are fifty to several hundred different genes, rather than just a few, that encode antibodies. These can be combined in different ways in B-cells to create some antibody diversity. Then a further stage of production of antibody variability occurs by more subtle changes in these genes within B-cells. The numbers of genes encoding antibodies and the mechanisms by which the genes are altered in B-cells account for the millions of different antibody molecules that can be produced.

In the midst of the research that was aimed at explaining the mechanisms of creation of antibody diversity and understanding what antibody molecules look like and how they function, Kohler and Milstein made their remarkable discovery of a method for making monoclonal antibodies.

Monoclonal antibody technology

The clonal selection theory for antibody production means that all of the antibody molecules produced by a single B-cell are identical and recognise and bind to a single antigen (or, rather, antigenic determinant). Such an antibody is called a **monoclonal antibody**. A **polyclonal antibody**, on the other hand, is really a mixture of antibody molecules that recognise different antigenic determinants. (It may also contain different antibodies that recognise the same antigenic determinant.) If an antigen is injected into an animal, many different antibodies to the antigen are frequently produced, some binding to the same region (antigenic determinant) of the antigen molecule, others binding to different regions. Serum from such an animal would therefore contain a mixture of antibodies to the same antigen: it would be a polyclonal antiserum.

Polyclonal antibodies are useful in many ways, but because they are mixtures of different antibody molecules, they pose certain problems to biochemists who study antibodies. With polyclonal antibodies, it is impossible to study the molecular properties of a single type of antibody molecule in isolation from the others. Ideally, one would wish to look at one antibody that is produced by a single immune cell (that is, a monoclonal antibody), rather than a mixture of antibodies produced by many different immune cells. Before 1975 the only major source of monoclonal antibodies were **myeloma cells**. Myelomas are cancers of the B-cells of the immune system. It is well known that when a cancer arises it is almost always derived from a *single* normal cell whose growth control mechanisms have gone haywire. A cancer of the B-cells, therefore, consists of cells derived from a single antibody-producing cell. Since each antibody-producing cell makes antibody molecules all of which are identical, having the same antigen specificity, then the antibody produced by a myeloma is a monoclonal antibody. Myelomas are rapidly growing cancers and can be isolated from the rest of the (normal) antibody-producing cells of an animal and grown in a test-tube. Myelomas were therefore valuable for studies aimed at understanding the basic structure and properties of antibody molecules. However, because myelomas, like all cancers, arise randomly within a normal B-cell population, any one of the millions of antibody-producing cells in the body could become a myeloma. The monoclonal antibodies produced by myelomas therefore tend to lack any known antigen specificity: the chances of identifying specific antigens that they would recognise are very slim, and one would need to screen hundreds of thousands of different antigens to be sure to identify their specificities. This is impracticable.

What was needed was a method for producing large amounts of monoclonal antibodies to known antigens at will. This would allow detailed studies of how antibodies bind to their specific antigens to be carried out, and it would eliminate any complications caused by the mixtures of antibodies that occur in polyclonal antibody preparations. B-cells are, unfortunately, very difficult to grow outside the body. If they are transferred from their bodily environment to special nutrient broths they usually die, in contrast to myeloma cells, which can readily be grown in isolation from the body. However, Kohler and Milstein hit upon a way of solving this problem and allowing B-cells to grow and produce their monoclonal antibodies to defined antigens in test-tubes.

Kohler obtained his doctoral degree from the University of Freibourg in Germany in 1984, after which he immediately joined Milstein's research group in Cambridge to carry out his postdoctoral research project. Milstein's team had been working for many years on the mechanisms by which B-cells generate antibody diversity. One approach that they were using at the time Kohler joined the laboratory was to fuse two different myeloma cells together to produce a hybrid cell. Such hybrid cells, like their unfused myeloma parent cells, were able to grow in test-tubes in the presence of appropriate nutrients. When pairs of myelomas were fused, the hybrids were found to express the specific antibodies of both parent myeloma cells. From these studies of myeloma hybrid cells, Milstein and his colleagues learned more about the mechanisms of antibody diversification in B-cells.

Kohler began to work on the antibodies produced by myeloma cells and on hybrids between two different myelomas, but he and Milstein realised that these studies would be much more informative if they could obtain a myeloma that produces an antibody to a known specific antigen. Unfortunately, the antigens recognised by the antibodies produced by almost all of the myelomas studied at that time had not been identified: they were monoclonal antibodies without any known antigen identity. A few myelomas were available whose antigens had been found, by chance, but these proved not to be useful to Kohler and Milstein, since they failed to grow well under laboratory conditions.

Notwithstanding these problems, the two scientists continued with their studies, and one approach they considered was that of screening many antigens with some of the myeloma-derived monoclonal antibodies, in order to determine whether or not they could find, by a process equivalent to searching for a needle in a haystack, the particular antigens that these antibodies recognised. This would be a laborious task, since many hundreds of thousands of antigens would need to be screened to have a reasonable chance of finding the antigens in question. Kohler and Milstein had the simple idea of fusing B-cells, obtained from an animal immunised with a particular antigen, with myeloma cells. If these hybrid cells could be obtained and grown in culture, they thought, it might be that they would produce not only the monoclonal antibody of the myeloma parent cell (against an unspecified antigen), but also the monoclonal antibody from the B-cell parent (against the specified antigen). In other words, if an antigen is injected into an animal, that animal will produce antibodies to the antigen. If the B-cells from these animals could be grown as hybrids with myeloma cells, then some of these hybrids should produce antibodies to the injected antigen, and they could readily be detected by their ability to recognise the antigen that was injected. B-cells fail to grow outside of an animal, but they might grow if they were fused to myeloma cells, which often grow prolifically in test-tubes.

Theoretical considerations suggested that the chances of production of specific monoclonal antibodies by B-cell/myeloma hybrids by the method proposed by Kohler and Milstein were slim and that the work involved would be very time-consuming. However, Milstein and Kohler decided to go ahead anyway, just in case their theoretical considerations were not entirely correct. Their perseverance was rewarded: when B-cells obtained from the spleens of mice injected with a chosen antigen were fused with myeloma cells, not only did the B-cell/myeloma hybrid cells grow well in test-tubes, but also some of the hybrid cells were found to produce antibodies specific to the antigen that was injected. Subsequent developments allowed such B-cell/myeloma hybrids, which are called **hybridomas**, to be obtained at ease. Hybridomas can now be obtained that produce antibodies to virtually any antigen. Kohler and Milstein had developed a much-needed method for making monoclonal antibodies at will to a chosen antigen.

An outline of the method for monoclonal antibody production is shown in Figure 34. An animal is injected with the antigen against which monoclonal antibodies are required. The animal produces antibodies to this antigen, and the antibodies can be detected in the serum of these animals. When the animal is producing sufficiently high levels of serum (polyclonal) antibodies to the antigen, its spleen is removed and its spleen cells, which are

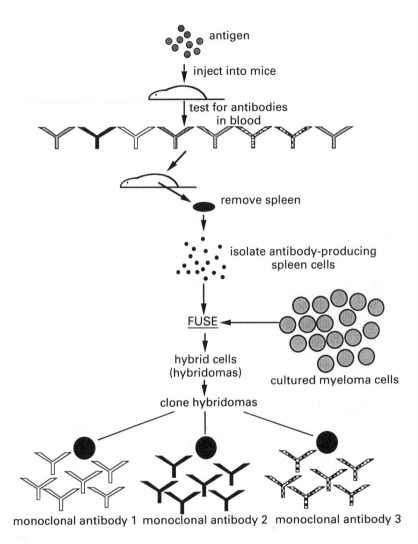

Figure 34. Method for monoclonal antibody production. Spleen cells from immunised mice are fused with myeloma cells and the resulting hybridomas are cloned. Each hybridoma (represented by a black circle) secretes a single and unique monoclonal antibody. Antibodies are represented as Y-shaped structures.

a rich source of the antibody-producing B-cells, are fused with myeloma cells (grown in test-tubes). Parent spleen B-cells fail to survive because they cannot grow in the nutrient medium, and special drugs are used to selectively kill the parent myeloma cells that do not fuse. In this way, only hybrid B-cell/myeloma cells (hybridomas) survive. The myelomas used nowadays do not produce their own antibody, and so the resultant hybridomas produce only the specific antibody that was made by the spleen B-cell partner. Hybridomas are separated from each other and allowed to multiply and grow in a special nutrient medium. This produces many **clones** of hybridomas; each clone consists of a population of identical cells derived from a single original hybridoma cell. Hybridoma clones are then screened for their ability to produce antibodies that recognise the injected antigen. Those that do make such antibodies are derived from a single B-cell in the animal that was injected with antigen: they produce monoclonal antibodies to the defined antigen. The spleen cell contributes the specific antibody producing property to the hybridoma, whereas the myeloma cell provides the important feature of being able to multiply and grow indefinitely in nutrient media in isolation from an animal's body.

Hybridomas producing monoclonal antibodies to a particular antigen can be grown in special nutrients in test-tubes forever, if that were necessary. They can also be frozen, stored for many years, then thawed and revived, when they continue to produce their antibodies. This means that monoclonal antibodies can be produced in limitless quantities, allowing their continued use in a standardised way. Polyclonal antisera, in addition to containing many different antibodies, are not available in limitless supply, since they are usually obtained from the blood of an animal and when the animal dies the supply of the polyclonal antibodies becomes exhausted. It is very difficult to produce exactly the same polyclonal antiserum by injecting another animal with the same antigen, and this means that polyclonal antibodies cannot be standardised as readily as monoclonal antibodies.

Since Kohler and Milstein made their discovery, monoclonal antibody technology has advanced in many ways. For example, two monoclonal antibodies can be joined together to produce an antibody that recognises not one, but two antigens. Another approach that is being taken to exploit the method is that of linking monoclonal antibodies to poisons. Some poisons, such as ricin, which is a protein obtained from castor beans, are very potent in their ability to kill living cells. It would be highly desirable in some medical circumstances to have a poison as powerful as ricin. For example, in order to kill cancer cells, a very potent toxin would be ideal. Unfortunately, ricin kills normal cells as well as cancer cells, so it is of no use by itself as an anti-cancer drug. However, the ricin molecule is made up from two different subunits: one of them (the B subunit) binds to living cells, whilst the other (the A subunit) enters cells and kills them. Without the A subunit, the B subunit binds to cells but causes them no harm. Without the B subunit, the A subunit will not bind to cells and cannot enter them. The actual killing property of ricin therefore lies with the A subunit, which is completely harmless on its own. Nevertheless, when the A subunit is joined to an antibody that binds to living cells, the A subunit will enter those cells and kill them almost as powerfully as it kills them when the A subunit is joined to the B subunit of ricin. In other words, when the B subunit of ricin is replaced with an antibody, the new antibody–A subunit complex is a potent poison. However, unlike the ricin B subunit, which binds to most cell types, antibodies can be produced that bind to specific kinds of cells and this means that antibodies can direct the A subunit to specific types of cell and selectively kill them. If now one obtains a monoclonal antibody that binds to an antigen on cancer cells, and if this antigen does not occur on normal cells, then the antibody–A subunit complex will kill cancer cells and not normal cells (Figure 35). The exciting possibility of destroying specific living cells, whilst leaving intact those cells one desires, is an area of intense research and one of its benefits may be the development of 'magic bullets', or antibody–poison conjugates, in which the antibody seeks out antigens on the surface of specific cancer cells (or other cells), and the poison kills those cells.

Diagnoses of diseases and other conditions, such as pregnancy, are also promising applications of monoclonal antibodies. Their high specificity for individual antigens, their long-term availability

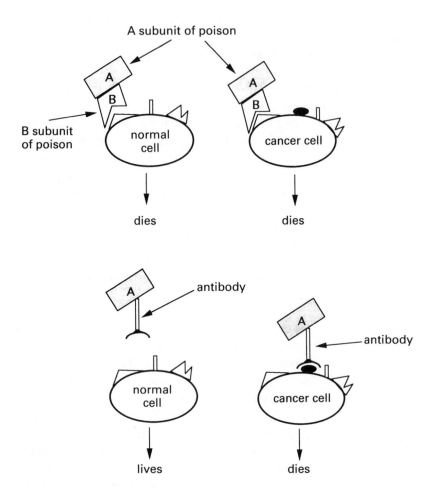

Figure 35. 'Magic bullets' produced by joining monoclonal antibodies to poisonous substances. Some poisonous proteins contain two subunits, A and B (top): the B subunit binds to cells whilst the A subunit then enters the cells and kills them. These poisons usually kill cancer and normal cells equally. If a monoclonal antibody was obtained that bound to a molecule specifically on the surface of cancer, but not normal, cells, and this antibody was joined to the poison's A subunit, the antibody–poison complex would bind to and kill only cancer cells.

and their ease of production and standardisation, make them well suited for diagnosis. If a disease or physiological situation is associated with changes in levels of a specific antigen, monoclonal antibodies that bind to the antigen can potentially be used to detect these altered levels and therefore to aid diagnosis. Pregnancy, for example, results in increased levels of the hormone, chorionic gonadotropin, in urine. If a plastic stick, which has antibody that binds to the hormone attached to it, is dipped into the urine, the hormone will stick to the antibody. The amount of hormone antigen 'captured' by the antibody on the stick can be determined by linking the process to a colour-producing reaction. For instance, in some commercial kits the stick turns blue, indicating that chorionic gonadotropin levels in urine are high and pregnancy is confirmed. Monoclonal antibodies have opened up new avenues and increased reproducibility and accuracy of such diagnostic tests.

Monoclonal antibodies will be used increasingly in biomedical research, medical diagnosis and disease treatment, and many more applications will undoubtedly be forthcoming.

Further reading

1. THE FATHER OF ELECTRICITY

Asimov, I. (1984) *The History of Physics*, Walker and Company, New York.

Encyclopaedia Britannica (1989) Faraday. 15th edition **19** pp. 83–6. Encyclopaedia Britannica, Inc., Chicago.

Faraday, M. (1855) *Experimental Researches in Electricity*, Great Books of the Western World **45** (1952) published by William Benton/Encyclopaedia Britannica, Inc., Chicago.

Gooding, D. and James F. A. J. L., eds. (1985) *Faraday Rediscovered: Essays on the Life and Work of Michael Faraday, 1791–1867*. Macmillan Press Ltd, Basingstoke, England.

Harré, R. (1983) *Great Scientific Experiments*, pp. 177–84. Oxford University Press.

King, R. (1973) *Michael Faraday of the Royal Institution*, The Royal Institution of Great Britain, London.

National Museum of Science and Industry (1991) *Michael Faraday – Celebrating 200 Years*, Swindon Press Ltd, Swindon.

Shamos, M. H. (1959) *Great Experiments in Physics: Firsthand Accounts from Galileo to Einstein*, pp. 128–58. Dover Publications, Inc., New York.

2. ONE GIANT LEAP FOR MANKIND

Bolton, W. (1974) *Patterns in Physics*, McGraw-Hill, Blue Ridge Summit, Pennsylvania.

Maxwell, J. C. (1873) *A Treatise on Electricity and Magnetism*, **1** and **2** (1954), Dover Publications Inc., New York.

Shamos, M. H. (1987) *Great Experiments in Physics*, Dover
Publications Inc., New York.

3. MEDICINE'S MARVELLOUS RAYS

Feinberg, J. G. (1952) *The Atom Story*, Allan Wingate, London.
Shamos, M. H. (1987) *Great Experiments in Physics*, Dover
Publications, Inc., New York.

4. THINGS THAT GLOW IN THE DARK

Curie, E. (1942) *Madame Curie*, published by the The Reprint
Society, London.
Encyclopaedia Britannica (1989) Atoms: Their Structure,
Properties and Component Particles. 15th edition **14** pp. 329–
80. Encyclopaedia Britannica, Inc., Chicago.
Shamos, M. H. (1987) *Great Experiments in Physics*, Dover
Publications, Inc., New York.

5. PARCELS OF LIGHT

Asimov, I. (1984) *The History of Physics*, Walker and Company,
New York.
Hoffmann, B. (1959) *The Strange Story of the Quantum*, Dover
Publications, Inc., New York.

6. DR EINSTEIN'S FOUNTAIN PEN

Asimov, I. (1984) *The History of Physics*, Walker and Company,
New York.
Bernstein, J. (1991) *Einstein*, Fontana Press, New York.
Gribbin, J. (1991) *In Search of the Big Bang: Quantum Physics
and Cosmology*, Corgi Books, London.
Kaufmann, W. J. (1985) *Universe*, W. H. Freeman and Co. Ltd,
New York.

Rothman, M. A. (1966) *The Laws of Physics*, Penguin Books, Middlesex, England.
Will, C. M. (1986) *Was Einstein Right?* Basic Books, New York.

7. THE BIG BANG, OR HOW IT ALL BEGAN

Boslough, J. (1990) *Beyond the Black Hole: Stephen Hawking's Universe*, Fontana, New York.
Gribbin, J. (1991) *In Search of the Big Bang: Quantum Physics and Cosmology*, Corgi Books, London.
Halliwell, J. J. (1991) Quantum Cosmology and the Creation of the Universe. *Scientific American*, December 1991, pp. 28–35.
Horgan, J. (1990) Universal Truths. *Scientific American*, October 1990, pp. 74–83.

8. MOLECULAR SOCCERBALLS

Curl, R. F. and Smalley, R. E. (1991) Fullerenes. *Scientific American*, October 1991, pp. 32–41.
Kroto, H. (1988) Space, Stars, C_{60} and Soot. *Science* **242** pp. 1139–45.
Kroto, H. (1992) C_{60}: Buckminsterfullerene, the Celestial Sphere that Fell to Earth. *Angewandte Chemie* **31** pp. 111–29.

9. JOSTLING PLATES, VOLCANOES AND EARTHQUAKES

Encyclopaedia Britannica (1989) Plate Tectonics. 15th edition **25** pp. 871–80. Encyclopaedia Britannica Inc., Chicago.
Erikson, J. (1988) *Volcanoes and Earthquakes*, TAB books, McGraw-Hill, Blue Ridge Summit, Pennsylvania.
Mason, S. F. (1991) *Chemical Evolution: Origins of the Elements, Molecules and Living Systems*, Clarendon Press, Oxford.
Van Andel, T. H. (1991) *New Views on an Old Planet: Continental Drift and the History of the Earth*, Cambridge University Press.

10. SODA WATER, PHLOGISTON AND LAVOISIER'S OXYGEN

Lavoisier, A. L. (1789) *Elements of Chemistry*, translated by R. Kerr, Great Books of the Western World **45** (1952), published by William Benton/Encyclopaedia Britannica, Inc., Chicago.

Leicester, H. M. (1971) *The Historical Background of Chemistry*, Dover Publications, Inc., New York.

McKie, D. (1980) *Antoine Lavoisier: Scientist, Economist, Social Reformer*, Da Capo Press, Inc., New York.

Orange, A. D. (1974) *Joseph Priestley*, Lifelines 31, Shire Publications Ltd, Aylesbury, Buckinghamshire, UK.

Priestley, J. (1970) *Autobiography*, Adams and Dart, Bath, England.

11. OF BEER, VINEGAR, MILK, SILK AND GERMS

Brock, T. D. (1961) *Milestones in Microbiology*, Prentice Hall, London.

Dubos, R. (1960) *Louis Pasteur, Free Lance of Science*, Da Capo Press, Inc., New York.

Vallery-Radot, R. (1937) *The Life of Pasteur*, Sun Dial Press, Inc., Garden City, New York.

12. OF MILKMAIDS, CHICKENS AND MAD DOGS

Dubos, R. (1960) *Louis Pasteur: Free Lance of Science*, Da Capo Press, Inc., New York.

Fisher, R. B. (1990) *Edward Jenner, 1749–1823*, Andre Deutsch Press, London.

Vallery-Radot, R. (1937) *The Life of Pasteur*, Sun Dial Press, Inc., Garden City, New York.

13. MALARIA'S CUNNING SEEDS

Major, R. H. (1978) *Classic Descriptions of Disease*, 3rd edition, Charles C. Thomas, Springfield, Illinois.

Megroz, R. L. (1931) *Ronald Ross: Discoverer and Creator*, Allen and Unwin, London.

14. PENICILLIN FROM PURE PURSUITS

Clark, R. W. (1985) *The Life of Ernst Chain: Penicillin and Beyond*, Weidenfield and Nicolson, London.

MacFarlane, G. (1980) *Howard Florey: The Making of a Great Scientist*, Oxford University Press.

MacFarlane, G. (1984) *Alexander Fleming: The Man and the Myth*, Harvard University Press, Cambridge, Massachusetts.

Malkin, J. (1981) *Sir Alexander Fleming: Man of Penicillin*, Alloway Publishing, Ayr.

Wilson, D. (1976) *Penicillin in Perspective*, Faber and Faber Ltd, London.

15. DNA, THE ALPHABET OF LIFE

Avery, O. T., MacLeod, C. M., and McCarty, M. (1944) Studies on the chemical nature of the substance inducing transformation of pneumococcal types. Induction of transformation by a deoxyribonucleic acid fraction isolated from pneumococcus type III. *Journal of Experimental Medicine* **79** pp. 137–58.

Judson, H. F. (1979) *The Eighth Day of Creation: Makers of the Revolution in Biology*, Jonathan Cape Ltd, London.

McCarty, M. (1980) Reminiscences of the early days of transformation. *Annual Review of Biochemistry* **49** pp. 1–15.

McCarty, M. (1985) *The Transforming Principle: Discovering that Genes Are Made of DNA*, W. W. Norton & Co. Ltd, New York and London.

16. CUTTING DNA WITH MOLECULAR SCISSORS

Boyer, H. W. (1971) DNA restriction and modification mechanisms in bacteria. *Annual Review of Microbiology* **25** pp. 153–76.

Kessler, C. and Manta, V. (1990) Specificity of restriction endonucleases and DNA modification methyltransferases – a review. *Gene* **92** pp. 1–248.

Nathans, D. and Smith, H. O. (1975) Restriction endonucleases in the analysis and restructuring of DNA molecules. *Annual Review of Biochemistry* **44** pp. 273–93.

17. DNA, THE MOLECULAR DETECTIVE

Burke, T., Doll, G., Jeffreys, A. J. and Wolff, R. (1991) *DNA Fingerprinting: Approaches and Applications*, Birkhauser Verlag, Basel, Switzerland.

Jeffreys, A. J. (1991) Advances in forensic science: Applications and implications of DNA testing. *Science in Parliament* **48** (1) pp. 2–7.

Jeffreys, A. J., Wilson, V. and Thein, S. L. (1985) Individual-specific 'fingerprints' of human DNA. *Nature* **316** pp. 76–9.

Yaxley, R. (1989) DNA fingerprinting. *International Industrial Biotechnology* **9** pp. 5–9.

18. MAGIC BULLETS

Milstein, C. (1980) Monoclonal antibodies. *Scientific American* **243** (4) pp. 66–74.

Milstein, C. (1987) Inspiration from diversity in the immune system. *New Scientist* 21 May 1987, pp. 55–8.

Winter, G. and Milstein, C. (1991) Man-made antibodies. *Nature* **349** pp. 293–9.

Yelton, D. E. and Scharff, M. D. (1981) Monoclonal antibodies: A powerful tool in biology and medicine. *Annual Review of Biochemistry* **50** pp. 657–87.

Index